OPTIMA FOR ANIMALS

R. McNeill Alexander

Optima for Animals

REVISED EDITION

PRINCETON UNIVERSITY PRESS, PRINCETON, NEW JERSEY

Copyright © 1996 by Princeton University Press
Published by Princeton University Press, 41 William Street,
Princeton, New Jersey 08540
In the United Kingdom: Princeton University Press, Chichester,
West Sussex

Originally published as *Optima for Animals* by
Edward Arnold (Publishers) Limited, 1982

Library of Congress Cataloging-in-Publication Data

Alexander, R. McNeill.
Optima for Animals / R. McNeill Alexander.—Rev. ed.
p.   cm.
Includes bibliographical references and index.
ISBN 0-691-02799-4 (cloth : alk. paper). ISBN 0-691-02798-6
(pbk. : alk. paper)
1. Evolution (Biology)—Mathematical models.   2. Mathematical
optimization.   I. Title.
QH371.A45   1996
591'.01'51—dc20   96-12454

This book has been composed in Lucida Bright

Princeton University Press books are printed on
acid-free paper and meet the guidelines for
permanence and durability of the Committee
on Production Guidelines for Book Longevity
of the Council on Library Resources

Printed in the United States of America
by Princeton Academic Press

10 9 8 7 6 5 4 3 2 1

10 9 8 7 6 5 4 3 2 1
(Pbk.)

Contents

Preface

EVOLUTION by natural selection favours those animals that are best adapted to their environment, whether in structure or in behaviour. Optimization theory is the branch of mathematics concerned with finding those structures and behaviours that are in some sense the best possible, and is therefore a very appropriate tool for trying to discover why animals have evolved in particular ways. It has been used most often in studies of feeding and reproduction, but has also been applied in many other fields ranging from anatomy to sociobiology. This book is about applications of optimization theory in all branches of zoology.

This is the second edition of a book which was designed principally for undergraduate biologists, but seems also to have been used and enjoyed by teachers and research workers. It assumes very little prior mathematical knowledge and I have tried to make the mathematical explanations clear and simple. You will not become an expert in optimal control theory, catastrophe theory, the theory of games and so on, simply by reading this book, but you should obtain some understanding of them and their uses.

The first chapter explains a few basic mathematical techniques and can be largely skipped by mathematically sophisticated readers. The next four chapters are the main section of the book. They describe and discuss applications of optimization theory to a wide range of zoological problems. Chapter 6 uses some of these examples in a discussion of the dangers and difficulties of the approach. Finally, Chapter 7 is (I hope) a convenient summary of all the optimization techniques used in the book, whether they were introduced in Chapter 1 or later. There is a short bibliography.

I have not tried to be comprehensive, by trying to include either all branches of optimization theory or all their applications in zoology. Instead I have tried to make the book short, interesting and useful.

In revising this book I have added information from many books and papers that have appeared since the first edition, but I have felt no need to substitute new examples for old ones if the latter seemed equally good.

November 1995   *R. McNeill Alexander*
Leeds

OPTIMA FOR ANIMALS

1

Introduction

1.1 Optimization and evolution

Evolution is directed by natural selection. Those sets of genes which enable animals to survive and reproduce best are most likely to be transmitted to subsequent generations. The ability to survive and reproduce can be measured by the quantity known as '*fitness*'. If a particular set of genes is possessed by n_1 members of the current generation and n_2 members of the next generation (the counts being made at the same stage of the life history in each generation), the fitness of that set of genes is n_2/n_1.

Evolution favours genotypes of high fitness but it does not generally increase fitness in the species as a whole. The reason is that fitness depends on the competitors which have to be faced, as well as on other features of the environment. A genotype which has high fitness now may have much lower fitness at some time in the future when a new, improved genotype has become common. In due course it will probably be eliminated, as evolution proceeds.

Though fitness itself may not increase, other qualities which affect fitness tend to improve in the course of evolution. For instance, natural selection generally favours characters that make animals use food energy more efficiently, enabling them to survive better when food is scarce and to divert more energy to reproduction when food is more plentiful. Natural selection generally favours characters which enable animals to collect food faster, so that they can either collect more food or devote more time to other activities such as reproduction. Natural selection generally favours characters that enable animals to hide or escape from predators more effectively. It favours characters that in these and other ways fit the animal best for life in its present environment.

A shopper looking for the best buy chooses the cheapest article among several of equal quality, or the best among several of equal price. Similarly, natural selection favours sets of genes which minimize costs or maximize benefits. The costs can often be identified as mortality or energy losses, the benefits as fecundity or energy gains.

Optimization is the process of minimizing costs or maximizing benefits, or obtaining the best possible compromise between the two. Evolution by natural selection is a process of optimization. Learning can also be an optimizing process: the animal discovers the most effective technique for some purpose by trial and error. Subsequent chapters show some of the ways in which optima are approached, in the structure and lives of animals. They are therefore concerned largely with maxima and minima. The rest of this chapter is about maxima and minima and ways in which they can be found. It introduces, as simply as possible, some of the mathematical concepts and techniques that are applied to zoological problems in later chapters.

1.2 Maxima and minima

In Fig. 1-1(a), y has a maximum value when $x = x_{max}$. In Fig. 1-1(b), y has a minimum value when $x = x_{min}$. It is easy enough to see where the maximum and minimum are when the graphs are plotted but it will be convenient to have a method for finding the maxima and minima of algebraic expressions, without drawing graphs. The most generally useful method is supplied by the branch of mathematics called differential calculus. The rest of this section can be skipped by readers who already know a little calculus.

The method of finding maxima and minima depends on the study of gradients. Figure 1-1(c) shows a straight line passing through the points (x_1, y_1) and (x_2, y_2). The gradient (slope) of this line is $(y_2 - y_1)/(x_2 - x_1)$. In this case $y_2 > y_1$ and $x_2 > x_1$ so the gradient is positive. In the case illustrated in Fig. 1-1(d), however, $y_2 < y_1$ and the gradient is negative.

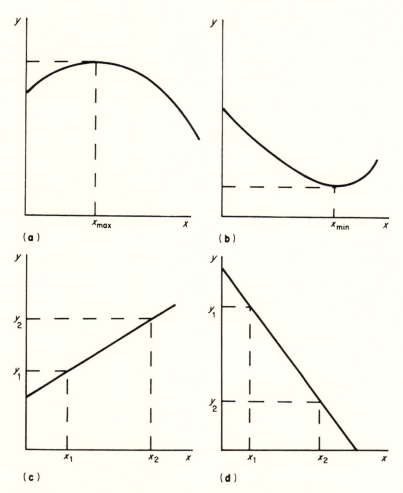

Figure 1-1. Graphs illustrating (**a**) a maximum; (**b**) a minimum; (**c**) a positive gradient and (**d**) a negative gradient.

A straight line has the same gradient all along its length. A curve can be thought of as a chain of very short straight lines of different gradients, joined end to end, so different parts of a curve have different gradients. In Fig. 1-1(a) the gradient is positive before the maximum (i.e. at lower values of x), zero at the maximum and negative after the maximum. Similarly in Fig. 1-1(b) the gradient is

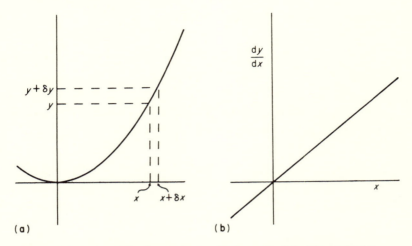

Figure 1-2. (a) A graph of y against x. **(b)** A graph of dy/dx against x. In both cases, $y = x^2$.

negative before the minimum, zero at it and positive after it. At a maximum the gradient is zero and decreasing, but at a minimum it is zero and increasing.

An example will show how gradients can be calculated. Figure 1-2(a) is a graph of $y = x^2$. It shows that y has just one minimum and no maximum, and that the minimum occurs when $x = 0$. Consider two points very close together on the graph, (x, y) and $(x + \delta x, y + \delta y)$. The symbol δx means a small increase in x and δy means the corresponding increase in y. The gradient of a straight line joining the two points is $\delta y/\delta x$. The smaller δx is, the more nearly is $\delta y/\delta x$ equal to the gradient of the curve at (x, y), which is represented by the symbol dy/dx. If δx is infinitesimally small, $\delta y/\delta x = dy/dx$. Note that dy/dx should be read as a single symbol: it is not a quantity dy divided by a quantity dx, still less can it be interpreted as $(d \times y) \div (d \times x)$.

In this particular case $y = x^2$

and also
$$y + \delta y = (x + \delta x)^2$$
$$= x^2 + 2x.\delta x + (\delta x)^2$$

Subtracting the first equation from the second

$$\delta y = 2x.\delta x + (\delta x)^2$$
$$\delta y/\delta x = 2x + \delta x$$

If δx is infinitesimally small, the term δx on the right hand side of the equation can be neglected, and $\delta y/\delta x = dy/dx$, so

$$dy/dx = 2x$$

It can be shown by similar arguments that if k is a constant

when $y = kx$	$dy/dx = k$	(1.1)
when $y = kx^2$	$dy/dx = 2kx$	(1.2)
when $y = kx^3$	$dy/dx = 3kx^2$	(1.3)
when $y = k/x$	$dy/dx = -k/x^2$	(1.4)
and when $y = k$	$dy/dx = 0$	(1.5)

All these cases are summarized by the general statement

$$\text{when } y = kx^n \qquad dy/dx = nkx^{n-1} \qquad (1.6)$$

Similar arguments can also be used to obtain expressions for the gradients of other functions of x, for instance

when $y = k.\log_e x$	$dy/dx = k/x$	(1.7)
and when $y = \exp(kx)^*$	$dy/dx = k.\exp(kx) = ky$	(1.8)

Notice particularly that in this last case, dy/dx is proportional to y. If x represented time this case would represent exponential growth, in which rate of growth is proportional to present size.

Other examples can be found in books on calculus. The process of obtaining dy/dx from y is called differentiation.

* $\exp(kx)$ means e^{kx}; $e = 2.718$

More complicated expressions are often easy to differentiate. Let y and z be two different functions of x. Then

$$\text{if } u = y + z \qquad du/dx = dy/dx + dz/dx \qquad (1.9)$$

$$\text{if } u = yz \qquad \frac{du}{dx} = z \cdot \frac{dy}{dx} + y \cdot \frac{dz}{dx} \qquad (1.10)$$

$$\text{and if } u = y/z \qquad \frac{du}{dx} = \left(z \cdot \frac{dy}{dx} - y \cdot \frac{dz}{dx} \right) / z^2 \qquad (1.11)$$

Also, if u is a function of y which in turn is a function of x

$$\frac{du}{dx} = \frac{du}{dy} \cdot \frac{dy}{dx} \qquad (1.12)$$

For instance to differentiate $u = (1 + x^2)^{\frac{1}{2}}$, write $y = 1 + x^2$. Then $du/dy = d(y^{\frac{1}{2}})/dy = \frac{1}{2} y^{-\frac{1}{2}}$ and $dy/dx = 2x$, so

$$du/dx = xy^{-\frac{1}{2}} = x/(1 + x^2)^{\frac{1}{2}}$$

Equations *(1.1)* to *(1.12)* enable us to discover values of x for which dy/dx is zero, in particular cases. Thus they help us to find maxima and minima. To distinguish between maxima and minima we also need to know whether dy/dx is increasing or decreasing. In other words, we need to know whether the gradient of a graph of dy/dx against x is positive (as in Fig. 1-2b) or negative, at the appropriate value of x. This can be discovered by differentiating again.

Since the symbol dy/dx was used for the gradient of a graph of y against x, it would be logical to use $d(dy/dx)/dx$ for the gradient of a graph of dy/dx against x. It is customary to write instead, for brevity, d^2y/dx^2.

At a minimum $dy/dx = 0$ and d^2y/dx^2 is positive.*   *(1.13)*

At a maximum $dy/dx = 0$ and d^2y/dx^2 is negative.*   *(1.14)*

These rules will be applied to the case $y = x^2$. Differentiation gives $dy/dx = 2x$, which is zero when $x = 0$. A second differentiation (using equation 1.1) gives $d^2y/dx^2 = 2$, which is positive

*There are some exceptional cases of maxima and minima at which d^2y/dx^2 is zero. See section 7.1.

when $x = 0$ (and for all other values of x). Hence y has a minimum when $x = 0$.

Most of the mathematical functions discussed in this book have just one maximum and no minima, or just one minimum and no maxima. Many other functions have a maximum and a minimum, or several of each. In such cases the rules *(1.13)* and *(1.14)* are generally adequate to find all the maxima and minima.

1.3 Optima for aircraft

The method which has just been explained will be illustrated by a simple example. It is about aeroplanes, not animals, but the same equation will be applied to birds in chapter 3. The power P required to propel an aeroplane in level flight at constant velocity u is given by the equation

$$P = Au^3 + BL^2/u \tag{1.15}$$

where A and B are constants for the particular aircraft and L is the lift, the upward aerodynamic force which supports the weight of the aircraft. The lift is produced by the wings deflecting air downwards, and the term BL^2/u represents the power required for this purpose. It is called induced power. The term Au^3 represents power which would be needed to drive the aircraft through the air, even if no lift were required. It is called profile power. As the speed u increases, profile power increases but induced power decreases, and there is a particular speed at which the total power P is a minimum. It can be found by differential calculus.

Differentiate equation *(1.15)*, using equations *(1.3)* and *(1.4.)*

$$dP/du = 3Au^2 - BL^2/u^2 \tag{1.16}$$

which is zero when u has the value $u_{\min P}$ given by

$$3Au_{\min P}^2 = BL^2/u_{\min P}^2$$
$$u_{\min P} = (BL^2/3A)^{\frac{1}{4}} \tag{1.17}$$

Differentiate equation *(1.16)*

$$d^2P/du^2 = 6Au + 2BL^2/u^3$$

which is positive for all positive values of u. This confirms that the power is a minimum, at the speed given by equation *(1.17)*.

This is the speed at which least power is needed to propel the aircraft, but it may not be the optimum speed. If the aircraft flew faster it would need more power but it would reach its destination sooner and might use less fuel for the journey. The energy E required to travel unit distance is given by

$$E = P/u$$
$$= Au^2 + BL^2/u^2$$

Differentiation gives

$$dE/du = 2Au - 2BL^2/u^3$$

and E has its minimum value at the speed $u_{\min E}$ given by

$$u_{\min E} = (BL^2/A)^{\frac{1}{4}} \qquad\qquad (1.18)$$

This is 32% faster than the speed $u_{\min P}$. A pilot wishing to remain airborne for as long as possible without re-fuelling should fly at $u_{\min P}$ but a pilot wishing to fly as far as possible without re-fuelling should fly at $u_{\min E}$.

1.4 Fitting lines

Figure 1-3(a) is a graph of a quantity y that depends on two variables, a and b. Since a, b and y all have to be represented, this has to be a three-dimensional graph. The third dimension is represented by the contours which give values of y. These contours show that y has a minimum value when $a = 2$ and $b = 3$.

Section 1.2 explained how maxima and minima can be found for functions of one variable. The rule for functions of two variables

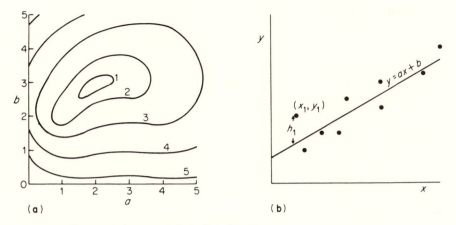

Figure 1-3. These graphs are explained in the text.

is very similar. At a maximum or minimum

$$\partial y / \partial a = 0 \text{ and } \partial y / \partial b = 0 \qquad (1.19)$$

Remember that dy/dx means the gradient of a graph of y against x. The symbol $\partial y/\partial a$ means the gradient of a graph of y against a, with b (and any other variables) held constant. Similarly $\partial y/\partial b$ means the gradient of a graph of y against b with all other variables held constant. A point at which equations *(1.19)* both hold may be a maximum, or a minimum, or neither. The rules for deciding which are more complicated than conditions *(1.13)* and *(1.14)* and are explained in books about calculus.

The usefulness of conditions *(1.19)* will be illustrated by explaining a standard statistical procedure. Experiments often lead to graphs like Fig. 1-3(b). The points are scattered (due perhaps to experimental errors) but suggest a sloping line. What straight line fits them best? This is a problem in optimization.

Any straight line can be represented by an equation

$$y = ax + b \qquad (1.20)$$

where a and b are constants (a is the gradient and b is the intercept). The problem is to find the best values of a and b, to fit the points (x_1, y_1), (x_2, y_2) etc. Suppose that particular values have been chosen. The first point is a height h_1 above or below the line, the second is h_2 above or below the line and so on. The standard method of choosing the best line is to choose the values of a and b which minimize the function

$$\Phi = h_1^2 + h_2^2 + h_3^2 + \ldots \ldots h_n^2$$

This can also be written

$$\Phi = \sum_{i=1}^{n} h_i^2$$

which means exactly the same thing. In this equation h_i means any of the heights h_1, h_2 etc. The symbol Σ means sum (add together). The letters above and below it show that all the h_i^2 have to be added together, from the first $(i = 1)$ to the last $(i = n)$.

The y coordinate of the first point is y_1 but the equation *(1.20)* suggests it should be (ax_1+b). Thus the height h_1 is (y_1-ax_1-b) and the function to be minimized is

$$\Phi = \sum_{i=1}^{n} (y_i - ax_i - b)^2 \qquad (1.21)$$

A graph of Φ against a and b would look rather like Fig. 1-3(a). Differentiating Φ with respect to a and b gives

$$\partial\Phi/\partial a = \sum_{i=1}^{n} [-2x_i(y_i - ax_i - b)]$$

$$\partial\Phi/\partial b = \sum_{i=1}^{n} [-2(y_i - ax_i - b)]$$

(The rule for differentiating a function of a function, equation *1.12*, was used). At the minimum, these must both be zero

$$\left.\begin{array}{l} \displaystyle\sum_{i=1}^{n} [-2x_i(y_i - ax_i - b)] = 0 \\[2em] \displaystyle\sum_{i=1}^{n} [-2(y_i - ax_i - b)] = 0 \end{array}\right\} \qquad (1.22)$$

The optimum values of a and b can be found, in any particular case, by solving (1.22) as a pair of simultaneous equations. This is the standard statistical technique of least-squares regression. Further details are explained in books on statistics.

If you have reached this point in the book, you should now understand what maxima and minima are, for functions of one or two variables. You should know the basic rules of differential calculus and understand how they can be used to find maxima and minima. This is sufficient mathematical preparation for the first half of the book, and the next two sections are intended to help you with the second half (particularly sections 4.6, 5.2, 5.3 and 5.4). I suggest that you read them now, but in case you prefer to go immediately to chapter 2 I have inserted references back to these sections, later in the book.

1.5 The best shape for cans

A cylindrical can is to be made to contain a volume V of food. What is the best shape (the best ratio of height h to diameter D)? It will be shown that two different approaches lead to the same answer.

It will be assumed that the best shape is the one that requires the least area of tinplate. The top and bottom of the can are circular discs, each of area $\pi D^2/4$. The circumference of the can is πD so the area of the rectangle of metal needed for its sides is πDh. The total area of metal needed is $(\pi D^2/2) + \pi Dh$. The volume of the can is $\pi D^2 h/4$. The problem can thus be stated

$$\text{minimize } \Phi = (\pi D^2/2) + \pi Dh$$
$$\text{subject to } \Psi = (\pi D^2 h/4) - V = 0 \tag{1.23}$$

Notice that this has the form, 'minimize the function Φ subject to the constraint that the function Ψ is zero'. Problems like this often arise and can be solved in several different ways.

Here is an obvious method that can be used in this case, but is difficult or impossible for some problems. The constraint gives

$$h = 4V/\pi D^2$$

Substitution of this in the expression for Φ gives

$$\Phi = (\pi D^2/2) + (4V/D)$$

and differentiation gives

$$d\Phi/dD = \pi D - 4V/D^2$$
$$= \pi D - \pi h$$

(using the constraint again). When Φ has its minimum value, $d\Phi/dD$ must be zero and so h must equal D. To make a can of given volume from least metal, the height and diameter should be made equal.

There is another method for getting the same result, which is sometimes more convenient. It will be presented here simply as a recipe, but explanations of why it works can be found in books on optimization (for instance Koo, 1977). Define a new function

$$L = \Phi + \lambda\Psi \tag{1.24}$$

The symbol λ represents a constant called a Lagrange multiplier. It can be shown that when D and h have the values which represent the solution to the problem, $\partial L/\partial D$ and $\partial L/\partial h$ must both be zero. In this particular problem

$$L = (\pi D^2/2) + \pi Dh + \lambda[(\pi D^2 h/4) - V] \tag{1.25}$$

so that at the minimum

$$\partial L/\partial D = \pi D + \pi h + \lambda[(\pi Dh/2) - 0] = 0 \tag{1.26}$$

and $$\partial L/\partial h = 0 + \pi D + \lambda[(\pi D^2/4) - 0] = 0 \tag{1.27}$$

Equation (1.27) requires either $D = 0$ (which is impossible, if the can is to hold anything) or $\lambda = -4/D$. By putting this value of

λ in equation *(1.26)* we get

$$\pi D + \pi h - 2\pi h = 0$$

$$h = D$$

so this method gives the same answer as the other one. The height should be made equal to the diameter.

Most of the food cans in my store cupboard have heights greater than their diameters. Their manufacturers were plainly not minimizing the area of metal used, but they may have been applying some other optimization criterion.

1.6 The shortest path

The problems of the aeroplane and of the line of best fit were answered by finding the optimum values for variables. Some problems require instead, the discovery of an optimum function. Consider, for instance, the problem of finding the shortest path between two points. The required answer is not a number but an equation, the equation of the shortest line that joins them. If we did not happen to know already that the answer is a straight line we would not know what mathematical form the equation should take, whether it should be $y = ax + b$ or $y = ax^b$ or $y = a^{bx}$ or something more complicated. Problems like this can sometimes be solved by a method called the calculus of variations.

Figure 1-4(a) shows a line (any line) joining the points P and Q. Consider a short segment of the line, from the point (x, y) to $(x + \delta x, y + \delta y)$. The length of this segment is (by Pythagoras) $[(\delta x)^2 + (\delta y)^2]^{\frac{1}{2}}$. Since δx and δy are short, $\delta y / \delta x$ is almost exactly equal to the gradient dy/dx so δy is approximately $(dy/dx).\delta x$. The length δl of the segment is given by

$$\delta l = [1 + (dy/dx)^2]^{\frac{1}{2}}.\delta x$$

To make subsequent equations less clumsy the symbol y' will be used to represent dy/dx.

$$\delta l = (1 + y'^2)^{\frac{1}{2}}.\delta x \qquad (1.28)$$

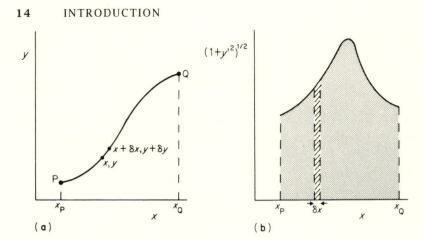

Figure 1-4. These graphs are explained in the text.

Figure 1-4(b) is a graph of $(1 + y'^2)^{\frac{1}{2}}$ against x. The narrow hatched strip under it is $(1 + y'^2)^{\frac{1}{2}}$ high and δx wide, so its area gives the value of δl. Other strips could be drawn representing the lengths of all the other segments of the line. Their areas could be added together to obtain the total length l of the line from P to Q. Thus the area of the whole stippled region under the graph gives the length l of the line. This is expressed by writing

$$l = \int_{x_P}^{x_Q} (1 + y'^2)^{\frac{1}{2}}.dx \qquad (1.29)$$

The expression on the right hand side of the equation, representing the area under a graph, is called an integral. The symbols x_P and x_Q at the bottom and top of the integral sign say where the area starts and ends. Books on calculus show how integrals can be evaluated, but the technique of integration is not needed here. It is sufficient for readers to understand what integrals are.

The problem of finding the shortest line joining two points is the problem of finding the equation relating x and y that minimizes the length l. This is a problem of the kind that the calculus of variations is designed to solve.

The standard problem of the calculus of variations is to find the equation relating x and y that minimizes (or maximizes) some function Φ that has the form

$$\Phi = \int_{x_a}^{x_b} f(x, y, y').dx \qquad (1.30)$$

In this equation, $f(x, y, y')$ means some function of any or all of x, y and y'. The most important theorem of the calculus of variations is that any solution to the problem must satisfy the Euler equation

$$\frac{\partial f}{\partial y} - \frac{d}{dx}\left(\frac{\partial f}{\partial y'}\right) = 0 \qquad (1.31)$$

for all values of x between x_a and x_b. A proof of the theorem is given in books by Koo (1977) and others.

In the problem of finding the shortest line, the function f is $(1 + y'^2)^{\frac{1}{2}}$. Since y does not appear in this expression (but only y'), $\partial f/\partial y = 0$. The example following equation (1.12) shows that $\partial f/\partial y' = y'/(1 + y'^2)^{\frac{1}{2}}$. Thus the Euler equation says

$$\frac{d}{dx}\left(\frac{y'}{(1 + y'^2)^{\frac{1}{2}}}\right) = 0$$

for all values of x between x_P and x_Q. This can only be true if y' ($= dy/dx$) is a constant, independent of x. Thus the gradient of the shortest path from P to Q (Fig. 1-4a) is constant and that path is a straight line.

This may seem a long-winded way of proving the obvious, but the calculus of variations is useful for solving other, less easy, problems of optimization.

1.7 Conclusion

This chapter has introduced some of the mathematics that will be applied to zoological problems in the rest of the book. A few more mathematical ideas will be introduced and explained when the need for them arises. This chapter has also illustrated an important general point. It is not sensible to claim in general terms that (for instance) a particular speed is optimal for aircraft, or that a particular shape is optimal for cylindrical cans. It is essential to state the criteria for optimization, that the speed is to be chosen (for instance) to minimize the power requirement and the shape to minimize the area of tinplate.

2

Optimum structures

THIS CHAPTER is about bones, eyes, eggshells, ears and guts. It examines some of the details of their structure and tries to explain why they have evolved as they have. These particular examples have been chosen from the whole range of animal structures because they seem to be unusually simple: it seems possible to formulate quantitative explanations of the features which will be discussed, using only a few simple ideas and equations. Also, the examples have been chosen to illustrate some of the very different ways in which optima can arise.

2.1 Tubular bones

The long bones in the legs of mammals are hollow tubes, filled with marrow. They have presumably evolved that way because they have to resist forces which tend to bend them. Tubes are stiffer and harder to break by bending than solid rods of the same length and weight, which is why bicycles and scaffolding are made of tubes. Lightness is a particularly desirable property for leg bones because legs have to be accelerated and decelerated in every step. The less the mass of a leg, the less the work needed to accelerate it.

Forces at right angles to rods and tubes tend to bend them. In Fig. 2-1(a) the force F exerts a moment Fx about any point in the cross-section XX. This moment tends to bend the beam at XX, and is described as the bending moment in that cross-section. Bending moments act on leg bones whenever forces that are not in line with the bones act on the feet. An example is shown in Fig. 2-1(b).

Fig. 2-1(c) represents a section through a tubular bone of radius r. Its marrow cavity has radius kr. The factor k could have any

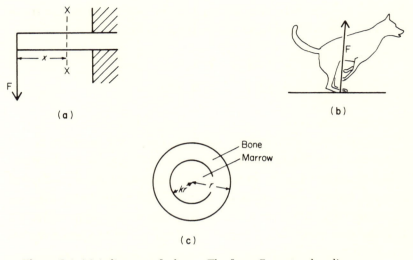

Figure 2-1. (a) A diagram of a beam. The force F exerts a bending moment Fx about the cross section XX. (b) A dog jumping with a force F acting on each hind foot. The drawing is based on a film and a force record, made simultaneously (Alexander, 1974). (c) A cross-section of a tube, to explain the parameter k.

value from 0 (for a solid bone) to very nearly 1 (for a bone with a very thin wall). Is any particular value of k the best?

Suppose a bone with a particular value of k is to be made just strong enough to withstand a bending moment M. Assume at first that the cross-section remains circular when the bone is bent: this assumption is justified unless k is very nearly 1. By a standard equation from engineering textbooks the bone must be given a radius r where

$$r = [M/K(1 - k^4)]^{\frac{1}{3}} \qquad (2.1)$$

(Alexander, 1983, pp. 127–131. The constant K is proportional to the strength of the material.) This equation shows that if k is large, r must also be large: a thin-walled bone needs a larger radius than if its wall were thicker. How does this affect the mass of the bone?

Since the external and internal radii of the bone in Fig. 2-1(c) are r, kr, the area of bone in the cross-section is $\pi r^2 - \pi k^2 r^2 =$

$\pi r^2(1-k^2)$. Let ρ be the density of bone and let m be the mass per unit length of the shaft of the bone. Then

$$m = \pi r^2 \rho(1-k^2) \qquad (2.2)$$
$$= \pi\rho(1-k^2)[M/K(1-k^4)]^{2/3}$$

(from equation 2.1).

The line 'bone only' in Fig. 2-2(a) has been calculated from this equation. It seems to show that the thinner the all of the bone (the larger the value of k) the lighter the bone can be made, but account has to be taken of the marrow. The cross-sectional area of the marrow is $\pi r^2 k^2$. The density of marrow is approximately half the density of bone. Hence the mass m' of marrow per unit length of bone is given by

$$m' = \frac{1}{2}\pi r^2 k^2 \rho$$
$$= \frac{1}{2}\pi k^2 \rho[M/K(1-k^4)]^{2/3} \qquad (2.3)$$

The line 'marrow' in Fig. 2-2(a) has been calculated from this equation. Notice that as k approaches 1, the mass of marrow increases very rapidly indeed. The line 'bone + marrow' has been obtained by adding m and m'. It represents the total mass of the bone and has a minimum when $k = 0.63$. This minimum is a shallow one. A marrow-filled bone with $k = 0.63$ is only 10% lighter than a solid bone of the same strength, and only very slightly lighter than equivalent bones with $k = 0.4$ or 0.7. Nevertheless we might expect to find that bones have evolved to have $k \simeq 0.6$. Table 2-1 shows some measured values. Most of them are less than 0.6, but all of them lie in the range 0.4 to 0.7.

The minimum in Fig. 2-2(a) could have been found by calculus as explained in Chapter 1 instead of by plotting the graph.

Bones can be made lighter if they are filled with air instead of marrow. This arrangement has not evolved in mammals, but many bird bones are filled with air. The mass of the air is negligible so the masses of air-filled bones of equal strength but different values

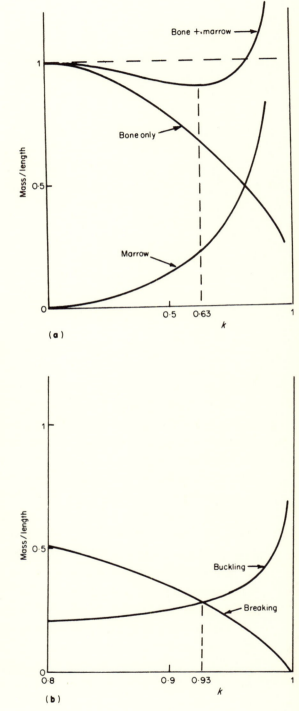

Figure 2-2. (a) Graphs of (mass/length) against the parameter k for bones which can withstand equal bending moments without breaking Mass/length is arbitrarily taken as 1 for a solid bone ($k = 0$). **(b)** Graphs showing the mass/length required for a air-filled bone, to prevent breaking and buckling under the same bending moment as in **(a)**. Note that the horizontal scale shows values of k between 0.8 and 1.0 only. The line for breaking is the 'bone only' line from **(a)**, modified slightly to take account of the flattening of the cross-section which occurs when tubes with high values of k are bent. The line for buckling has been calculated from equation (4.4) (p. 148) of Alexander (1983), using data on the mechanical properties of bone taken from Wainwright *et al.* (1975). The Young's modulus used was the value for transverse stretching.

Table 2.1. Values of the constant k (the ratio of the internal diameter to the external diameter), measured at the mid-points of the shafts of various mammal bones.

	Hare	Fox	Lion	Camel	Buffalo
Femur	0.57	0.63	0.56	0.62	0.54
Humerus	0.55	0.59	0.42	0.66	0.51

of k are given by the line 'bone only' in Fig. 2-2(a). This line seems to show that the greater the value of k, the lighter an air-filled bone can be made.

This is misleading. If k is made too large the bone will not break under an excessive bending moment. Instead, it will fail by buckling, like a plastic drinking straw. There is an equation which predicts the bending moment required to buckle a tube of given dimensions, made of material of given elastic properties (Alexander 1983). The equation works quite well for celluloid but may not be reliable for bone because the 'grain' of bone gives it different elastic properties in different directions. The equation has nevertheless been used to estimate the masses of air-filled bones, with different values of k, which just fail to buckle under the bending moment assumed in Fig. 2-2(a). These masses are shown in Fig. 2-2(b) (labelled 'buckling') together with the values for breaking.

To withstand the bending moment, the bone must neither break nor buckle. Hence the required mass in Fig. 2-2(b) is given by whichever of the two lines is the higher, at the particular value of k. The best value of k (the one which minimizes the required mass) is 0.93, where the two lines intersect. This estimate may be highly inaccurate because it is based on too simple a theory of buckling, but the general conclusion is clear: there is a (high) value of k which is the optimum for strength with lightness. The air-filled bones of birds have values of k much higher than the marrow-filled bones of mammals. The humerus of a swan has $k \simeq 0.9$ and has internal struts which must help to prevent buckling.

Beams and bones can sometimes be made lighter if they are tapered. In Fig. 2-1(a), the bending moment Fx is proportional to

x so it is less near the free end of the beam than near the fixed end. There is no need for the free end to be thick enough to withstand the bending moment at the fixed end. Similarly, bending moments change along the lengths of bones. The tibia, for instance, is held and moved by several muscles which attach to it near the end, but the only forces which normally act on its other end are those transmitted to it through the ankle joint. The knee end of the tibia is comparable to the fixed end of the beam in Fig. 2-1(a), and the ankle end to the free end. Any force at the ankle which is not in line with the tibia, sets up bending moments in the tibia. The bending moment at any point on the tibia (except between the muscle attachments and the knee) is proportional to the distance of that point from the ankle. This is the situation, for instance, in the jumping dog in Fig. 2-1(b).

Bending moments are generally the most important causes of stress in bones, except very near joints. Equation *2.1* shows that for tubes with the same value of k, the radius r is proportional to the cube root of the bending moment M needed to break them. If the tibia were a tube of circular section and constant k it should be tapered (for optimum strength with lightness) so that its radius was proportional to the cube root of the distance from the ankle.

Sections of different parts of the tibia have different shapes (Fig. 2-3), so a more complex analysis is needed. The ability of a cross-section to withstand bending moments is given by a quantity called the section modulus, which can be calculated from the dimensions of the section. Section modulus is (bending moment)/(maximum resulting stress). Figure 2-3 shows that for the tibia of a dog, it is very nearly proportional to distance from the ankle. The tibia is tapered very nearly in the ideal way, to make the maximum stress the same in every cross-section.

2.2 Strengths of bones

The preceding section looked for the most economical possible design, for a bone of specified strength. This section discusses how strong bones should be. Too weak a bone would be likely to

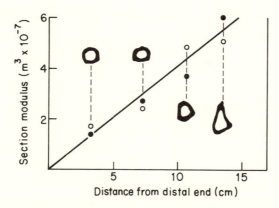

Figure 2-3. A graph of section modulus against distance from the distal end of the bone, for a dog tibia of overall length 17 cm. One of the points for each cross-section refers to tensile stresses and the other to compressive stresses. The measured cross-sections are also shown, with their anterior edges uppermost (Alexander, 1975).

break, with possibly disastrous consequences. Too strong a bone would contain an excessive amount of material and be unnecessarily cumbersome. This would be true, even if the bone were a tube of optimal proportions tapered in the optimal way. What is the optimum strength for a bone? This section is based on a paper by myself (Alexander, 1981).

Factor of safety is a concept used by engineers. It is the ratio of the intended strength of a structure, to the maximum load it is expected to have to bear. For instance, an engineer designing a beam to carry an expected maximum load of 10 tonnes may aim to make it strong enough to carry 20 tonnes, applying a factor of safety of two.

The leg bones of dogs probably have to withstand larger bending moments when the dog makes a big jump, than in any other activity, excluding accidents. These bending moments have been calculated from force records of dogs jumping (Fig. 2-1b). The strengths of the bones have been calculated from their dimensions. It has been calculated in this way that the humerus and tibia have factors of safety of about 3. Similar factors have been calculated for leg bones of several other mammals, running and jumping. These factors are rather unrealistic because larger bending moments are likely to act on the bones if the animal slips and falls.

Factors of safety are advisable (and required by engineers' codes of practice) because neither loads nor strengths can be predicted accurately. It may be expected that the maximum load imposed by traffic on a bridge will be 10 tonnes but the actual maximum load may be 15 tonnes, or only 7. The bridge may be designed to be capable of supporting 20 tonnes but it may actually be able to support 22 tonnes, or only 18. The maximum load and the strength are not precisely known but can be represented, in the light of experience, by probability distributions (Fig. 2-4a). It may be possible to estimate, for instance, that there is a probability of 0.001 that the maximum load will exceed 15 tonnes and a probability of 0.001 that the strength will be less than this. The overlap of the two probability distributions indicates the possibility that the load may exceed the strength and the structure will fail.

The hypothesis has been proposed, that evolution by natural selection tends to adjust factors of safety so as to minimize a total cost, which will be explained (Alexander, 1981). If a bone has a factor of safety s, let the probability that it will fail be $P(s)$. This symbol is used to indicate that the probability depends on (is a function of) the factor of safety. If the bone does fail the animal is hampered by lameness or some other disability, which can be represented by a cost F. The cost associated with the possibility of failure is $P(s).F$. Let the cost associated with growth and use of the bone be $U(s)$. This takes account of energy and materials used in growing the bone, energy used in moving it (for instance, in running) and any effect the mass of the bone has on the animal's athletic performance. The hypothesis is that natural selection tends to minimize the total cost $\Phi(s)$, given by

$$\Phi(s) = P(s).F + U(s) \qquad (2.4)$$

The costs should ideally be calculated in terms of fitness but it may be more convenient to use energy or some other currency.

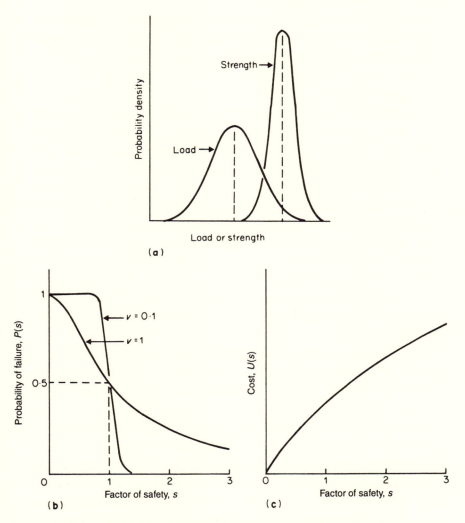

Figure 2-4. Schematic graphs illustrated the discussion of factors of safety.

Figure 2-4(b) shows schematic graphs of $P(s)$ against s. When $s = 0$, $P(s) = 1$; the bone will certainly fail. When $s = 1$, $P(s) = 0.5$ and when s is very large, $P(s)$ is very small. The graphs are based on the lognormal distribution, which resembles the more familiar normal probability distribution but is more appropriate in this case because it makes $P(s) = 1$ when $s = 0$. Figure 2-4(c) is a schematic graph of $U(s)$ against s. It has been assumed that $U(s)$ is proportional to the mass of the bone, which for a bone of given length is proportional to the 2/3 power of the maximum bending moment it can resist (equation 2.2).

Thus

$$U(s) = C.s^{2/3} \qquad\qquad (2.5)$$

where C is a constant.

Figure 2-5(a) to (d) shows four identical graphs of $U(s)$ against s. Also shown are graphs of $P(s).F$ against s, for four different values of F, and graphs of the total cost $\Phi(s)$ (equation 2.4). In Fig. 2-5(d), $\Phi(s)$ has a maximum at one value of s and a minimum at another (indicated by arrows). In (c) there are again a maximum and a minimum, but they are closer together. In (b) the maximum and minimum have merged to form a point of inflection, a horizontal step on the graph which is neither a maximum nor a minimum. In (a) no trace of the maximum or minimum remain; the gradient of the graph is positive throughout. Events like this, in which maxima and minima merge and disappear, are called bifurcations by mathematicians. The branch of mathematics that deals with their consequences is called catastrophe theory.

The hypothesis is that evolution tends to adjust s so as to minimize $\Phi(s)$. In Fig. 2-5(d) and (c), $\Phi(s)$ has minima at particular (non-zero) values of s, and bones with these values of s could be expected to evolve. In (a), however, there is no minimum and the lowest value of $\Phi(s)$ is obtained with $s = 0$, that is with no bone at all. In this case the bone should be eliminated by natural selection. A small change in the environment that reduced the cost of failure F only slightly, but happened to move it from one side of the bifurcation to the other, could result in a large (catastrophic) evolutionary change.

It is not clear whether the change should be expected to occur precisely when the minimum vanished. At stages between Fig. 2-5(b) and (c), $\Phi(s)$ is lower for $s = 0$ than at the minimum. The bone might be lost at this stage, if an evolutionary jump over the intervening maximum were possible. Alternatively, if evolu-

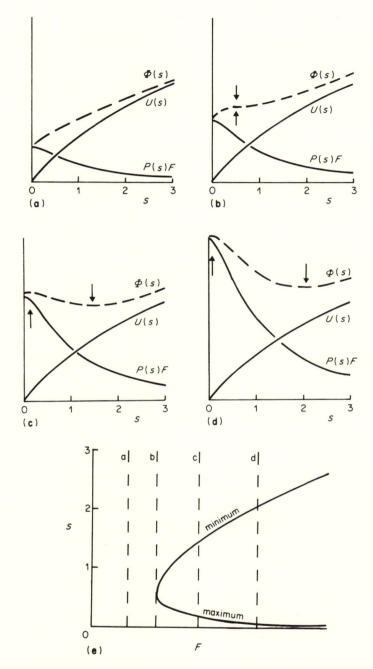

Figure 2-5. (a) to (d) Graphs of costs represented in equation 2.4 against factor of safety s, for four different values of the cost of failure F. (e) A graph showing the values of s that maximize or minimize the cost $\Phi(s)$, for different values of F.

tion could proceed only by small changes the bone would not be lost until stage (b), because while the minimum existed, any *small* movement from it would increase $\Phi(s)$. Catastrophe theory makes a distinction between systems that are capable of jumping over maxima to find the lowest of several minima (which are said to follow the Maxwell convention) and ones which remain trapped in a minimum for as long as it exists (which follow the perfect delay convention).

The conclusion, that a bone should be eliminated by natural selection if F falls too low, could have been reached by common sense. If breakage causes very little disadvantage, the bone has little value and it is a waste of energy and materials to grow it and carry it around.

Figure 2-5(e) shows the values of s that give $\Phi(s)$ its maximum and minimum values. This figure refers to one particular value of ν, the parameter that determines the shape of graphs of $P(s)$ against s (Fig. 2-4b). This parameter tells how variable load and strength are. If the probability distributions (Fig. 2-4a) are wide, ν is large, but if they are narrow ν is small. Figure 2-6(a) shows how the values of s required to minimize $\Phi(s)$ depend on ν as well as on F. This is a three-dimensional graph with contours to show optimum values of s. The vertically hatched area is beyond the bifurcation, where the bone should be eliminated even on the perfect delay convention. In the stippled region a local minimum remains, but elimination of the bone would be better.

An optimum factor of safety implies an optimum probability of failure, $P(s)$. These probabilities are shown in Fig. 2-6(b). They are low when variability ν is low and cost of failure is high, but they are high near the catastrophe.

The costs in equation 2.4 are difficult to measure, but probabilities of failure of bones can be estimated in a few cases. The Viverridae is a family of carnivorous mammals that includes the mongooses and genets. About 300 viverrid carcases were collected in East Africa, most of them by trapping. Their skeletons were examined, and 1% of the long bones of their legs were found to show evidence of having been broken at some time. (A healed fracture

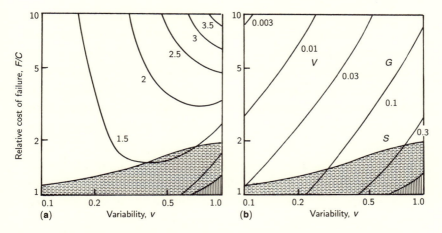

Figure 2-6. Graphs illustrating the theory of optimum factor of safety for bones. The horizontal axes give the variability parameter v. The vertical axes give the cost of failure F divided by the constant C of equation 2.5. The contours show (a) the optimum factor of safety and (b) the corresponding probability of failure. The letters in (b) mark points that are suggested tentatively for V, the leg bones of viverrids; G, the leg bones of gibbons and S, the antlers of stags. The shaded regions are explained in the text. (Modified from Alexander, 1981.)

generally leaves an obvious swelling on the bone, or the two halves of the healed bone may be misaligned. Fresh fractures that might have occurred in the trap were ignored.) If the animals had not been trapped they might have lived on average twice as long, so the probability that a particular leg bone will fail during a viverrid's life can be estimated as 0.02. This may be an underestimate because fatal accidents involving fractures were excluded from the sample, but no better estimate is available. Medical records indicate a similar probability of fracture for human leg and arm bones, but the probability of fracture for gibbon leg bones seems to be much higher, about 0.07. Gibbons feed high in trees, and are apt to fall. The elaborate antlers of red deer stags have an even higher probability of fracture. The stags wrestle with them, in contests for females (section 4.10). They are replaced annually, but a study on the Isle of Rhum (Scotland) showed that the probability that some branch of a particular antler would be broken in the course of a season was about 0.2. (This value may be unusually high. Investigators in other places have formed the impression that breakage is quite rare.)

The leg bones of viverrids and gibbons and the antlers of stags have presumably all evolved to have optimum factors of safety. Three points have been marked in Fig. 2-6(b), in positions where the observed probabilities of failure would be optimal. These positions are not unique, but they seem quite plausible. The bones of gibbons are liable to more variable loads than those of viverrids, because gibbons are apt to fall from unpredictable heights, but the cost of failure seems likely to be about the same for both. Hence gibbon bones probably have larger values of ν than viverrid ones, but similar values of F/C. The antlers of stags suffer variable loads because of the accidents of fighting, but costs of failure are probably less than for leg bones. Loss of small branches seems to have little effect, and though loss of a whole antler may reduce breeding success this season the antler will be replaced next season. Hence ν is probably high and F/C low.

This seems to show that existing data on bone fractures are not inconsistent with the hypothesis that has been expounded. It would be satisfying to have more data and a more rigorous analysis.

2.3 Compound eyes

The eyes of insects consist of many independent units called ommatidia, each pointing in a slightly different direction from its neighbours (Fig. 2-7a). Typical insects that are active by day have pigment between adjacent ommatidia, so light cannot cross from one ommatidium to another. Each ommatidium sees only in the direction of its own axis.

An eye of a given size may have a large number of small ommatidia or a smaller number of larger ones. Barlow (1952) and Snyder (1977) discussed the optimum size for ommatidia.

It seems clear from their structure that each ommatidium sees only a spot of light, not a detailed image. To see how this limits the

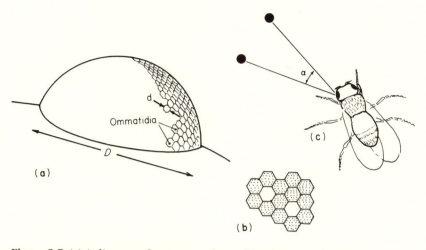

Figure 2-7. (a) A diagram of a compound eye. (b) A diagram of a group of ommatidia, only two of which are illuminated. (c) A diagram of an insect looking at two objects separated by an angle α.

detail that can be seen, think of an insect looking at two tiny bright objects against a dark background. These objects will not be seen as separate unless the light from them falls on separate ommatidia with an unilluminated ommatidium between them (Fig. 2-7b). This means that the angle α between the objects (Fig. 2-7c) must be at least twice the angle between the axes of adjacent ommatidia. If the eye is a hemisphere of diameter D with ommatidia of diameter d at their outer ends (Fig. 2-7a), the angle between adjacent ommatidia is $2d/D$ radians. Objects cannot be distinguished unless

$$\alpha \geqslant 4d/D \qquad (2.6)$$

This suggests that the smaller d is, the finer the detail that the eye can see. However, light travelling through a small aperture is spread out a little by diffraction. This sets a limit to the resolving power of compound eyes and other optical instruments (including microscopes). Light of wavelength λ passing through an aperture of diameter d cannot form separate images of objects separated by an angle less than about $1.2\lambda/d$ radians. Similarly, the unillu-

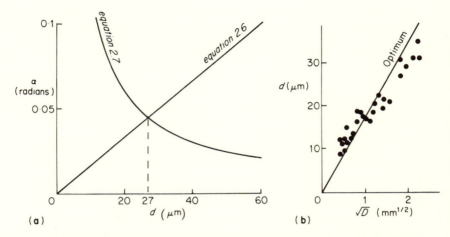

Figure 2-8. (a) Graphs of the angle α between objects which can be resolved by a compound eye of diameter 2.4 mm, against ommatidial diameter d. The two lines show the lower limits to α given by equations 2.6 and 2.7. (b) A graph of ommatidial diameter d against the square root of eye diameter D. The line shows the optimum values given by equation 2.8. The points are measured values for bees and wasps. Modified from Barlow (1952). The wavelength of light in the eye is assumed to be 0.5 μm in both (a) and (b).

minated ommatidium between the illuminated ones in Fig. 2-7(b) cannot exclude the light from the bright objects unless

$$\frac{1}{2}\alpha \geqslant 1.2\lambda/d \qquad (2.7)$$

(The axis of this ommatidium makes angles of $\frac{1}{2}\alpha$ with the directions of both objects.)

Figure 2-8(a) shows the results of calculations for light of wavelength 0.5 μm in a compound eye of diameter 2.4 mm. This is the size of the eyes of bees, *Apis mellifera*. For any given ommatidial diameter d, two objects cannot be distinguished unless the angle α between them lies above both lines. The optimum diameter which would enable a bee to see the finest detail is 27 μm, where the lines intersect. Barlow (1952) found that the diameters of bee ommatidia were actually a little less than this, about 21 μm.

At the intersection in Fig. 2-8(a), the ommatidial diameter has the optimal value d_{opt} and the limits to α set by equations 2.6 and 2.7 are equal. Hence for eyes of any diameter D

$$4d_{opt}/D = 2.4\lambda/d_{opt}$$
$$d_{opt} = (0.6\lambda D)^{\frac{1}{2}} \qquad (2.8)$$

The line in Fig. 2-8(b) represents this equation. It shows theoretical optimum ommatidial diameters for eyes of different sizes. The points show actual diameters for a selection of bees and wasps, ranging from a tiny chalcid wasp less than 1 mm long ($D = 0.2$ mm) to a giant tropical bee 60 mm long ($D = 5$ mm). The points lie quite close to the theoretical line, indicating that all these insects have ommatidial diameters quite close to the optima for their sizes of eye. A line fitted to the points by the method outlined in section 1.4 would, however, have a smaller gradient than the theoretical line.

This analysis contains a hidden assumption which is probably justified for most of the bees and wasps of Fig. 2-8(b), but would not be justified for some other insects. The assumption is, that the insects use their eyes largely in bright light. To understand the relevance of the assumption it is necessary to remember that light consists of discrete particles, the photons (Snyder, 1977).

Consider an insect in a simple environment consisting of patches of just two degrees of brightness. An ommatidium of area A looking at a bright patch receives photons from it at a mean rate NA (number per unit time) but one looking at a dim patch receives photons at a smaller rate nA. The ommatidia distinguish brightness by a method that consists in effect, of counting the photons received in a short time interval δt. This interval has to be short: otherwise moving objects would not be seen clearly. The mean number of photons received in this interval is $NA.\delta t$ from bright patches, but only $nA.\delta t$ from dim patches. The precise numbers received in particular intervals vary at random around these means, according to probability distributions like the ones shown in Fig. 2-9. The variabilities of the numbers are described by their standard deviations σ_N and σ_n. If the means are well

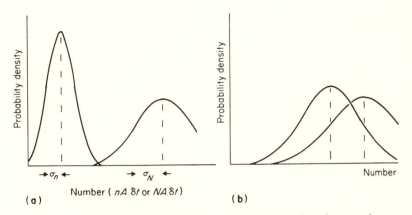

Figure 2-9. Schematic graphs representing probability distributions for the numbers of photons received in time δt by ommatidia looking at dim and bright patches of the environment.

separated as in Fig. 2-9(a), the ommatidia can distinguish reliably between bright and dim patches. If they are close together as in Fig. 2-9(b) they cannot distinguish reliably, because it is quite likely that in a particular interval an ommatidium looking at a bright patch would receive *fewer* photons than one looking at a dim patch.

It is advantageous for the insect to be able to distinguish dim patches from bright ones, even when n is not much less than N. For it to be able to do this, the standard deviations of the numbers of photons being counted should be as small as possible compared to the means: $\sigma_N/NA.\delta t$ should be as small as possible. A standard result from statistics says that if photons are emitted at random times, the standard deviation of the number received in an interval δt equals the square root of the mean. Thus $\sigma_N = (NA.\delta t)^{\frac{1}{2}}$ and the quantity to be made as small as possible is $(NA.\delta t)^{-\frac{1}{2}}$. In a bright environment, N is large and there is no difficulty in keeping this quantity small. In a dim environment, N is small and it may be advantageous to make ommatidial area A or counting time δt larger than would be appropriate in a bright environment. In particular, it may be advantageous to have larger ommatidia than equation *2.8* suggests as optimal. Snyder (1977) has worked out the theory in more detail.

Moths and many other insects that fly by night have eyes that allow light to pass from one ommatidium to another, so this theory does not apply to them. Dragonflies, however, have ommatidia separated by pigment. The dragonfly *Zyxomma*, which is active mainly at dusk, has much larger ommatidia than dragonflies that are active at midday.

Very small ommatidia would receive photons at a low rate even in bright light, so it would not be surprising if very small diurnal insects had larger ommatidia than equation *2.8* suggests. This helps to explain the discrepancy in Fig. 2-8(b), between the gradient suggested by the points and the gradient of the theoretical line.

2.4 Eggshells

The shells of birds' eggs are pierced by tiny pores which allow oxygen to diffuse in, but also allow water vapour to diffuse out. If the pores were too small the embryo would suffocate and if they were too large it would dry up. The need for oxygen is greatest at the end of incubation, when the embryo is respiring fastest. Let the partial pressure of oxygen in the egg at this stage be p_o less than its partial pressure in the atmosphere. Let the egg lose a fraction a of its initial mass by evaporation of water during incubation. Large values of p_o or a would be harmful.

The optima in previous sections were due to different factors affecting one quantity. For instance, the effects described by equations *2.6* and *2.7* both affect the resolving power of insect eyes. In this section we are concerned with the effects of pore size on two quantities of different kind, p_o and a.

How can an optimum be defined? The simplest likely hypothesis is that natural selection tends to minimize a function Φ given by

$$\Phi = a + K p_o \qquad (2.9)$$

It seems likely that given values of p_o and a would be equally harmful to the eggs of all species, so K will be assumed to be a constant.

Diffusion through an eggshell or any other porous barrier can be described by the equation

$$J = Gp \qquad (2.10)$$

Here J is the rate of diffusion of the gas, p is the difference between partial pressures of the gas on opposite sides of the barrier and G is a constant called the conductance of the barrier for the particular gas. In this discussion of eggshells, the values of G and p for oxygen and water vapour, respectively, will be written G_o, p_o and G_v, p_v.

As already defined, p_o refers to the end of incubation when the rate of consumption of oxygen has its maximum value of R. From equation *2.10*

$$R = G_o p_o \qquad (2.11)$$

An egg of mass m loses a mass am of water during the incubation period T so the rate of loss of water is am/T, and

$$am/T = G_v p_v \qquad (2.12)$$

(The value of p_v depends on the humidity of the atmosphere.)

Equations *2.11* and *2.12* give values for p_o and a which have been substituted in equation *2.9* to give

$$\Phi = (G_v p_v T/m) + (KR/G_o)$$

The oxygen and water vapour diffuse through the same pores so G_v/G_o is a constant and the equation can be written

$$\Phi = (G_v p_v T/m) + (K'R/G_v) \qquad (2.13)$$

(where $K' = (G_v/G_o)K$). If the conductance G_v is increased, the first term on the right hand side of the equation increases and the second decreases. There must be an optimum conductance, $G_{v,opt}$, which minimizes Φ.

Differentiating

$$\mathrm{d}\Phi/\mathrm{d}G_v = (p_v T/m) - (K'R/G_v^2)$$

At the minimum $\mathrm{d}\Phi/\mathrm{d}G_v = 0$ and

$$p_v T/m = K'R/G_{v,\mathrm{opt}}^2$$
$$G_{v,\mathrm{opt}} = \sqrt{K'Rm/p_v T} \qquad (2.14)$$

We cannot calculate $G_{v,\mathrm{opt}}$ for any egg because we do not know the value of K', but we can predict how it will depend on the mass m of the egg. There is a general rule that related animals of different sizes use oxygen at rates about proportional to (body mass)$^{0.75}$. Measurements of the oxygen consumption of a range of eggs, from the 1.3 g egg of a wren to the 170 g egg of a goose, give $R \propto m^{0.77}$, in good agreement with the rule (Rahn, Paganelli and Ar, 1974). Small eggs develop faster than large ones and it is found that $T \propto m^{0.22}$. (Note that these two proportionalities give $RT \propto m^{0.99}$, suggesting that eggs of different sizes metabolize almost equal fractions of their initial food stores during incubation.) Thus equation 2.14 gives

$$G_{v,\mathrm{opt}} \propto \sqrt{m^{0.77} m/p_v m^{0.22}} = m^{0.78} p_v^{-0.5} \qquad (2.15)$$

The conductance G_v has been measured for a range of eggs from wren to ostrich, by observing their rates of loss of weight in a desiccator. Figure 2-10 shows the results plotted on logarithmic coordinates. The line was fitted by the method outlined in section 1.4. Its gradient is 0.78, indicating $G_v \propto m^{0.78}$, which agrees with the theory. The points are scattered on either side of the line so it is impossible to say with confidence that the exponent is precisely 0.78, but statistical analysis of the data shows that the probability is 0.95 that the exponent lies between 0.73 and 0.83.

Grebes (*Podilymbus*) build floating nests of decaying vegetation which they leave for long periods during the day, covering the eggs with wet nest material. The decay of the vegetation keeps the eggs

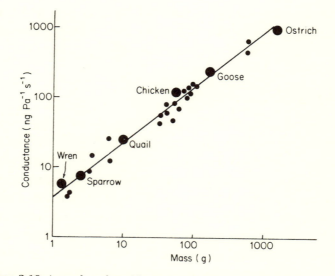

Figure 2-10. A graph on logarithmic coordinates of water vapour conductance against mass, for the eggs of various birds. Data from Ar, Paganelli, Reeves, Greene and Rahn (1974).

warm while the parents are away. These eggs have an exceptionally humid environment: p_v is therefore small, and equation *2.15* suggests that shell conductance should be unusually high. Davis, Platter-Reiger, and Ackerman (1984) found that the conductance of grebe eggs was 2.7 times higher than expected for eggs of their size.

The conductance is proportional to the total cross-sectional area A of the pores in the shell and inversely proportional to their length s: $G_v \propto A/s$. Geometrically similar bodies have lengths proportional to (volume)$^{1/3}$ and areas proportional to (volume)$^{2/3}$. Therefore if eggs of different sizes were geometrically similar in all respects, including the details of their pores, A would be proportional to $m^{2/3}$ and s to $m^{1/3}$. The conductance G_v would be proportional to $m^{1/3}$, which is very different from the observed proportionality to $m^{0.78}$. Measurements on a range of eggs show that s is actually proportional to $m^{0.46}$ and A to $m^{1.24}$: the shells of large eggs are relatively thicker than those of small eggs but are very much more porous, so that they have the required conductance. The pores occupy 0.02% of the area of a chicken eggshell, but 0.2% of the area of an ostrich eggshell.

2.5 Semicircular canals

The first section of this chapter showed that bones are constructed in extremely economical fashion. Both the ratio k (internal diameter/external diameter) and the taper of bones seem to have evolved to give the required strength with minimum weight. The nervous system seems very different. It includes many examples of several identical structures serving a function which apparently needs only one of them. This is called redundancy.

There is a striking example in the semicircular canals of the ear. Each canal is sensitive to rotation about a single axis, which deflects a jelly-like structure in the canal by an amount roughly proportional to the angular velocity. The only information to be transmitted to the brain is the amount of deflection. It could be signalled by the frequency of action potentials in a single axon (nerve fibre). The signal is actually carried by a very large number of axons. Wersäll (1956) counted 1200 axons in the nerve to one of the semicircular canals of the guinea pig. These axons are very small (most of them have diameters of 3–5 μm) but the arrangement nevertheless seems wasteful.

The next few paragraphs suggest possible advantages of redundancy. It increases reliability. Suppose each axon is liable to fail but can be repaired or replaced. Let the probability that it will be out of action at any instant be P. The probability that n identical axons will all be out of action simultaneously is P^n, which is less than P (since P must lie between 0 and 1). The probability of having at least one axon working, enabling the animal to get the information it requires, is increased by increasing n.

This may be why redundancy evolved in some cases, but it is not a plausible explanation for the 1200 axons to one semicircular canal. Suppose natural selection reduced P^n to 10^{-12}, implying failure of the system for one second every 3000 years. It seems exceedingly unlikely that it could do so, since selection cannot work effectively on very rare events. Even if it did, it would not require $n = 1200$ *unless* $P \simeq 0.98$: unless, that is, each axon worked for only 2% of the time. I do not believe that axons are so unreliable. A different explanation for the 1200 axons is needed.

Redundancy increases accuracy. Suppose that the signal in each axon is subject to random variation so that for the same stimulus, the signals on different occasions have a standard deviation σ. The mean of the signals from several identical axons varies less than the signal from a single axon, for the same stimulus, because the random errors of the individual axons tend to cancel each other out. The extent of the random variation of the mean is described by its standard error, which would be $\sigma/\sqrt{n}$ if n axons were in use. It follows, however, from an earlier paragraph that only $n(1 - P)$ axons are likely to be working so the standard error can be estimated as $\sigma/\sqrt{n(1 - P)}$. The standard error can be reduced, improving accuracy, by increasing n.

Energy and materials are needed to grow and maintain axons, so possession of n axons implies a cost proportional to n. This cost must be weighed against the benefits of improved accuracy. This suggests a hypothesis, like the hypothesis for eggshells expressed by equation 2.9. In the evolution of the nerve supplies of sense organs, natural selection may tend to minimize a function Φ given by

$$\Phi = [\sigma/\sqrt{n(1 - P)}] + Kn \qquad (2.16)$$

where K is a constant. This seems quite plausible and it is easy enough to find the minimum by calculus, but I have not been able to devise any quantitative test for the hypothesis.

2.6 Herbivore guts

Mammals all have guts built of the same basic components (oesophagus, stomach, large and small intestines) but there are wide variations in the sizes and functions of these parts. Carnivore guts differ from herbivore guts, and even among herbivores there is great variety of design. This section asks whether the variations can be explained as optimal designs for different diets.

Two distinct processes are used to break down plant food: digestion by the herbivore's own enzymes and fermentation by microorganisms living in the gut. The foodstuffs present inside the cells of plants (sugars, starch, proteins, and fats) can be digested or fermented, but cell wall materials such as cellulose can only be fermented. In fermentation, the microorganisms extract some of the food energy, using it for metabolism and growth. However, fatty acids remain as fermentation products that the microorganisms cannot use. By absorbing them, the herbivore can obtain about 75% of the food energy originally present in the cellulose.

Digestion proceeds fastest in relatively narrow tubes such as the small intestines of mammals, in which there is no mixing of newly arrived food with food that has been in the tube for some time. Fermentation is only possible in wider chambers whose contents can be well mixed: the microorganisms would soon be washed out of a narrow tube by the food travelling down the gut.

Herbivore guts consist of fermentation chambers and digestive tubes connected in series. In cows part of the stomach is enlarged as a huge fermentation chamber, the small intestine is a digestive tube, and there is a second, smaller fermentation chamber in the hindgut. Horses do not use the stomach for fermentation but have a large hindgut fermentation chamber.

There are advantages and disadvantages in placing fermentation chambers at each end of the gut. Foregut fermentation (as in cows) has the advantage that the microorganisms that proliferate in the foregut pass eventually to the intestine, where the cow can digest them; but it has the disadvantage that cell contents arrive initially in a fermentation chamber and get fermented, yielding less energy to the cow than if they had been digested. Hindgut fermentation has the converse advantages and disadvantages; cell contents arrive first in a digestive tube and get digested, but microbes pass out undigested in the faeces.

There seems to be no reliable way of predicting how the balance of advantage will fall, by purely verbal reasoning. Instead, a math-

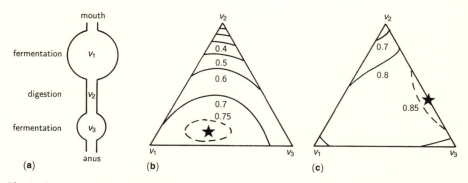

Figure 2-11. The effects of herbivore gut design on energy gain from various diets. (a) shows the gut model. In (b) and (c) the triangles represent the range of possible gut designs, with different values of v_1, v_2, and v_3. Contours show the proportion of the food's energy content that the herbivore gains. (b) refers to a poor diet with a low proportion of cell contents and (c) to a moderate diet. These graphs have been derived by a version of the model that takes account of the diminishing volume of the food as it travels along the gut. From Alexander (1993).

ematical model has been used (Alexander, 1991, 1993). Consider an animal with a gut of total volume V consisting of a foregut fermentation chamber of volume $v_1 V$, a digestive tube of volume $v_2 V$ and a second fermentation chamber of volume $v_3 V$, connected in that sequence (Fig. 2-11a). Let the animal eat food containing specific proportions of cell contents and cell wall, at a specified rate. Make realistic assumptions about the rate constants of digestion and fermentation and about the fractions of the energy of fermented food that are converted to microbial growth or to fatty acids. This has been done, and a computer program that follows the fate of the food as it travels down the gut, calculating how much energy the herbivore gains from it, has been written. The calculations are just like those used in engineering to predict the output of chemical reactors.

Some results are shown in Fig. 2-11. Each triangular graph compares the performance of different gut designs for dealing with a particular diet. Each point in the triangle represents a different gut design. The bottom left corner represents a gut consisting of a foregut fermentation chamber and nothing else ($v_1 = 1; v_2 = $

$v_3 = 0$); a point half way up the left edge represents a gut consisting of a foregut fermentation chamber and a digestive tube of equal volumes, and nothing else ($v_1 = v_2 = 0.5; v_3 = 0$); and points inside the triangle represent guts having all three components present. The contours show the fraction of the food energy that the herbivore is able to absorb and stars mark the gut designs that give maximum food absorption.

Fig. 2-11b refers to a poor diet—that is, to one that contains a high proportion of cell wall materials and only a small proportion of cell contents. The calculated optimum gut design has a large foregut fermentation chamber, a small digestive tube, and a second, smaller fermentation chamber ($v_1 = 0.6; v_2 = 0.1; v_3 = 0.3$). Cattle and antelopes have guts much like this and eat relatively poor diets of grass or leaves of shrubs.

Fig. 2-11c refers to a somewhat richer diet and shows an optimum design with no foregut fermentation chamber, but a large hindgut one ($v_1 = 0; v_2 = 0.4; v_3 = 0.6$). Horses have guts like this, which may seem puzzling, because many scientists consider them to be adapted for poor diets. In discussing this, we should of course consider the natural diets of wild horses, not the food given to domesticated ones. Large volumes of grass stems (very poor food) have been found in wild zebra guts, but these animals may have been getting most of their energy from grass seeds borne on the stems, in which cases their diet would have been a moderate one.

Simulations for very rich, plentiful diets predict an optimum gut consisting of a digestive tube only. Spider monkeys (*Ateles*) eat a rich diet of fruit, and their only fermentation chamber is a very small one in the hindgut.

3

Optimum movements

CHAPTER 2 was about structures but this chapter is about movements. It tries to explain some of the features of bird flight and of the motion of athletes, other mammals and tortoises.

3.1 Bounding flight

Very often, if you watch a small bird flying, you will see that it does not flap its wings all the time. For a few metres it flaps them, but then it folds them and keeps them folded for a few metres, before flapping them again (Fig. 3-1). This alternation between flapping and folding the wings makes the bird rise and fall as it flies, so the style of flying is called bounding flight. Tits and finches often use it. Why do they fly like this?

Flight requires a high power output from the muscles. (It has been shown for instance that budgerigars use oxygen about six times as fast when flying, as when perching.) Birds may learn to fly in the least exhausting way, or if styles of flight are inherited natural selection will favour styles which minimize energy costs. How does the energy cost of bounding flight compare with the cost of level flight with the wings beating continuously?

The energy cost of bounding flight has not been measured, but it can be estimated by a simple calculation using equation *1.15* (Lighthill, 1977). This equation gives the power P required to propel an aircraft flying at speed u. It is intended to refer to fixed-wing aircraft but can probably be applied without too much error to reasonably fast flapping flight. (In very slow flight the wings move much faster than the body and the equation does not apply.) The equation can be written

$$P = (A_b + A_w)u^3 + (BL^2/u) \qquad (3.1)$$

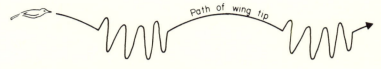

Figure 3-1. A diagram illustrating bounding flight.

A_b, A_w and B are constants for the particular bird and L is the lift. $(A_b + A_w)u^3$ is the profile power and (BL^2/u) is the induced power. (These components of power were explained in Chapter 1.) The constant A of equation 1.15 has been divided into two parts so that a distinction can be made between the part $A_b u^3$ of the profile power which is required to propel the bird's body through the air and the part $A_w u^3$ associated with the wings.

Let the bird flap its wings for a fraction a of the time. The mean lift must equal the weight mg of the bird so if there is no lift while the wings are folded, the lift while they are flapping must be mg/a. Hence the power required to propel the bird while the wings are flapping is

$$P_{\text{flap}} = (A_b + A_w)u^3 + (Bm^2g^2/a^2u) \qquad (3.2)$$

While the wings are folded flat against the body there is no induced drag, if there is no lift, and little more profile drag than if the wings had been removed completely from the body. Hence the power needed to propel the bird while its wings are folded is

$$P_{\text{fold}} = A_b u^3 \qquad (3.3)$$

The mean power over a whole cycle of bounding flight is

$$\begin{aligned} \overline{P} &= aP_{\text{flap}} + (1-a)P_{\text{fold}} \\ &= A_b u^3 + aA_w u^3 + (Bm^2g^2/au) \qquad (3.4) \end{aligned}$$

The wing muscles are doing no work while the wings are folded so the bird slows down (losing kinetic energy at a rate equal to P_{fold}). It speeds up again, gaining kinetic energy, while the wings are flapping. If the intervals between bursts of flapping are short (which they are in real life) the fluctuations of speed are small.

As a increases, one term in equation 3.4 increases and another decreases. There is a particular value of a which minimizes the power needed for flight and is presumably optimal. Differentiating equation 3.4 with respect to a,

$$\mathrm{d}\overline{P}/\mathrm{d}a = A_\mathrm{w}u^3 - (Bm^2g^2/a^2u)$$

The derivative $\mathrm{d}\overline{P}/\mathrm{d}a$ is zero when a has its optimal value a_opt given by

$$a_\mathrm{opt} = (B/A_\mathrm{w})^{\frac{1}{2}}(mg/u^2) \tag{3.5}$$

(A second differentiation would show that $\mathrm{d}^2\overline{P}/\mathrm{d}a^2$ is positive, confirming that the turning point is a minimum.)

For low speeds equation 3.5 gives $a_\mathrm{opt} > 1$, which is impossible, so if the bird flies slowly it should flap its wings all the time ($a = 1$). For higher speeds equation 3.5 implies that as speed increases, a_opt decreases: the faster the bird flies, the smaller the fraction of the time for which it should flap its wings.

That explanation for bounding flight seemed convincing until birds were observed bounding while hovering, with no forward speed at all. A new explanation was needed, and one was suggested (Rayner, 1985). In the discussion so far we have considered only the *mechanical* work that the bird must do, but what matters to the bird is the *metabolic* energy cost. If the conditions of bounding flight enable the muscles to operate more efficiently, they may use less metabolic energy, even if they are doing more work.

For any particular muscle, there is a rate of shortening at which it works most efficiently, needing the least metabolic energy to perform a required amount of work. Suppose that a bird's wing muscles are adapted to be most efficient during the most strenuous activities, such as rapid climbing flight. In less-strenuous activities it will have the choice of beating its wings more slowly (in which case the muscles will work less efficiently) or of alternating rests with bursts of beating at the most efficient rate. The second alternative, which is bounding flight, may be the more economical.

3.2 High jumping

This next example concerns athletics. It is mainly about high jumping, but a few words about pole vaulting will help to introduce the problem.

Pole vaulters run up for their jump as fast as they can, giving their bodies as much kinetic energy as possible. They plant the pole which bends, storing elastic strain energy. Then the pole recoils, throwing the athlete into the air, giving him or her potential energy. Kinetic energy is converted to elastic energy which in turn is converted to potential energy. The faster the athlete runs up, the higher the vault can be.

Accordingly, good pole vaulters and long jumpers run up to the take-off point at a fast sprinting speed, about 10 metres per second. However, high jumpers run up more slowly, at about 7 m/s. Could they not jump higher, if they ran faster?

Fig. 3-2a represents a simple computer model that was devised to tackle this question (Alexander, 1990). It has a rigid body and simple legs. Its only muscle is a knee extensor which has realistic properties: the force it can exert is reduced when it is shortening (extending the knee) and enhanced when it is being stretched (resisting bending of the knee [Fig. 3-2b]). At the end of its run-up, the model places its foot on the ground with the knee almost straight, and activates the muscles. The knee bends and extends again (as it does at take-off in real high- and long-jumping) and throws the athlete into the air. The computer program calculates the forces and movements of take-off, and the length and height of the jump. This was done for a range of run-up speeds and initial leg angles.

Figs. 3-2c and d show results. In each case the horizontal axis shows speed from a slow jog to a very fast sprint. The vertical axis shows a wide range of leg angles. In Fig. 3-2c the contours show the length of the jump. This is greatest when the athlete runs up

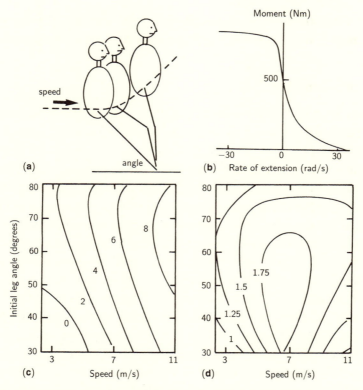

Figure 3-2. (a) A model of human jumping. **(b)** A graph showing the moment the muscle was assumed to be able to exert at different rates of knee extension or bending. **(c)** The lengths and **(d)** the heights of the model's jumps, in metres. These are shown as contours on plots of initial leg angle against run-up speed. From Alexander (1990).

as fast as possible, setting down the leg at about 65°, which is what good long jumpers do. In Fig. 3-2d the contours show the height of the jump. In this case the maximum is near the middle of the graph, at a speed of about 7 m/s and an angle of 50°. Good high jumpers take off like this. Though the model is so simple it has predicted well the different techniques of the two events.

Some of the parameters needed for the model (especially muscle properties) are not known accurately. To check that the good agreement between the model's predictions and real jumping was not just a lucky accident, I tried the effect of quite large changes of these parameters. The calculated optimum speeds and angles were changed relatively little.

3.3 Walking and running

People walk to go slowly and run to go faster. The change from walking to running is abrupt: there is no intermediate between the two gaits. The change is also predictable. Adult men of average height normally make the change at about 2 m s^{-1} although it is possible to walk faster, as athletes in walking races show. What is the difference between walking and running and why are they used at different speeds?

A stride, in walking or running, is a complete cycle of leg movements. The duty factor is the fraction of the duration of the stride, for which each foot is on the ground. In walking the duty factor is greater than 0.5 so there are stages of the stride when both feet are on the ground simultaneously (Fig. 3-3a, b, c). In running, the duty factor is less than 0.5 so there are stages when neither foot is on the ground (Fig. 3-3d).

Figure 3-3 shows the vertical components of force which act on the feet in walking and running. The measurements on which it is based were obtained by asking people to walk across a force platform, a force-sensitive panel set into the floor. Notice that the graphs of force against time have different shapes for the three speeds of walking and for running. In walking, the graph for each footfall has two maxima with a minimum between them, but in slow walking the minimum is shallow (Fig. 3-3a) and in fast walking it is deep (Fig. 3-3c). In running, there is just one maximum (Fig. 3-3d). (A small subsidiary maximum due to vibrations following impact with the ground is ignored.) I will try to explain the difference between speeds of walking, as well as the difference between walking and running.

The shapes of graphs of forces on feet, against time, have been described by a simple form of Fourier analysis (Alexander and Jayes, 1980). Figure 3-4(a) is half a cycle of a graph of cos θ against

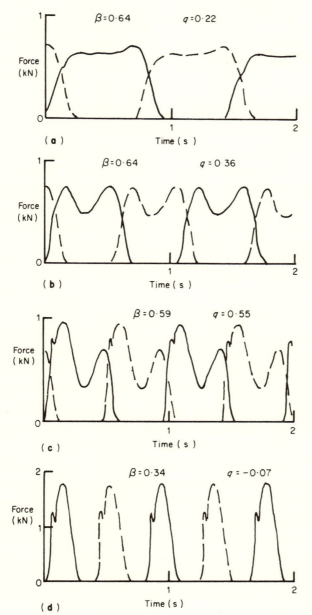

Figure 3-3. Forces exerted by the feet of a man walking at three different speeds (**a,b,c**) and running (**d**). The vertical components of the forces are plotted against time using continuous lines for one foot and broken ones for the other. The duty factors (β) and shape factors (q) are explained in the text. These graphs are based on the data of Alexander and Jayes (1980).

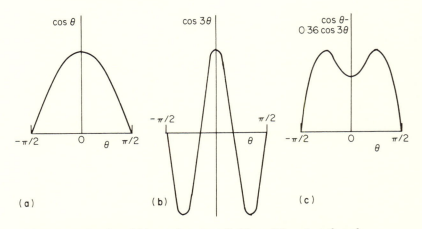

Figure 3-4. Graphs of (a), cos θ against θ; (b) cos 3θ against θ, and (c) (cos θ − 0.36 cos 3θ) against θ.

θ. It starts at zero (for $\theta = -\pi/2$), rises to a maximum and falls to zero again (at $\theta = \pi/2$). Figure 3-4(b) is a graph of cos 3θ against θ over the same interval. It too starts and ends at zero, but it has two minima and a maximum. Figure 3-4(c) is a graph of (cos θ − 0.36 cos 3θ) against θ. It has the same shape as the graphs of force against time, for individual footfalls, in Fig. 3-3(b). The other graphs in Fig. 3-3 can be imitated in the same way, by using other factors instead of 0.36.

This idea needs to be expressed more formally. Consider a foot which is set down on the ground at time $-T/2$ and lifted at time $T/2$. At time t in this interval let the vertical component F of the force exerted on it by the ground be given by

$$F = K[\cos(\pi t/T) - q\cos(3\pi t/T)] \qquad (3.6)$$

where K and q are constants for this particular step. (Notice that at the beginning and end of the step, when $t = \pm T/2$, $\pi t/T = \pm\pi/2$). This equation cannot describe exactly the patterns of force which act on people's feet, but it gives quite good approximations. The parameter q will be called the shape factor because it describes the shapes of graphs of force against time. Values for each speed of walking or running are shown in Fig. 3-3.

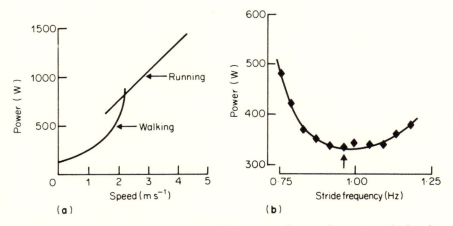

Figure 3-5. The metabolic power consumed by men walking and running, calculated from measurements of oxygen consumption. In (a) power is plotted against speed. The subjects used whatever stride frequency they preferred, at each speed. In (b) power is plotted against stride frequency for walking at 1.5 ms^{-1}. The arrow shows the preferred frequency. Data from (a) Margaria, 1976, (b) Zarrugh and Radcliffe (1978).

This figure shows that the following changes occur as a person speeds up from a slow walk to a run. While he or she is still walking the duty factor falls only slightly, and the shape factor increases gradually. When he or she changes to running, the duty factor and shape factor both fall abruptly. These are the changes which are to be explained.

Figure 3-5(a) suggests that the change from walking to running is made to minimize energy costs. The data were obtained by collecting and analysing the air breathed out by men walking and running at various speeds. The power being dissipated by metabolism was calculated from the rate of consumption of oxygen. The lines show that below 2.2 m s^{-1}, running requires more power than walking and (by extrapolation) that above 2.2 m s^{-1}, walking would require more power than running. The change of gait is made at about the speed at which walking becomes more expensive than running.

To discover whether the duty factors and shape factors which people use minimize energy costs, we need to know what the costs would be if they walked and ran differently. It might be possible to train people to use unusual duty factors and shape factors, and to measure their rates of oxygen consumption, but it would certainly be difficult. Mathematical modelling has been used instead (Alexander, 1992; Minetti & Alexander, 1995).

When a muscle shortens while exerting a force it does work, giving mechanical energy to the body. When a muscle is forcibly stretched it acts like a brake, converting mechanical energy to heat. Both processes use metabolic energy and the model takes account of both. It calculates the metabolic energy consumed in a stride, for any given duty factor and shape factor.

Results from the model are shown in Fig. 3-6, in which each graph refers to a different speed. These graphs show how metabolic energy cost depends on both duty factor and shape factor. Energy cost is shown by contour lines, so the regions where it has minima look like valleys on a map.

In Fig. 3-6a, representing a very low speed, there are two minima. The deeper of the two (called the global minimum, marked by a filled star) is at a duty factor of 0.55 and a shape factor of 0.4, representing a walk. There is also a less-deep minimum (a local minimum, marked by a hollow star) at a duty factor of 0.2 and a shape factor of 0.1, representing an exceedingly slow jog. To keep the energy cost as low as possible, the global-minimum walk should be used, but the local-minimum jog is better than any other running gait. At a moderate walking speed (Fig. 3-6b) the global optimum has moved to the highest possible shape factor. Shape factors greater than 1.00 are impossible because they would require the foot to exert a negative force on the ground (i.e., to pull upward) at the mid-point of its period on the ground. At 2 m/s (the speed at which we change from walking to running, Fig. 3-6c) running has become the more economical gait, and the global minimum is at a duty factor of 0.25 and a shape factor of 0.1. It remains close to that position as speed increases further (Fig. 3-6d).

Thus Fig. 3-6 indicates that people should walk at speeds below 2 m/s and run at higher speeds. In the walking range, shape factor

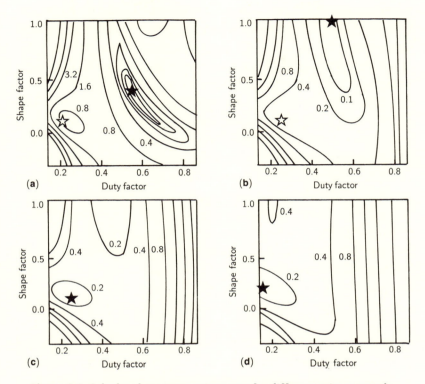

Figure 3-6. Calculated energy requirements for different gaits at speeds of (**a**) 0.4 m/s; (**b**) 1.2 m/s; (**c**) 2.0 m/s and (**d**) 3.2 m/s. The axes represent duty factor and shape factor and the contours show energy cost in joules per metre, per newton of body weight. From the data of Minetti & Alexander (1995).

should increase as speed increases, until the abrupt change to running is made. We do change our gait like this as we change speed (see Fig. 3-3), but though the simple theory predicts the trends correctly it does not give accurate predictions of duty factors and shape factors for particular speeds.

The consequences of using different values of the shape factor q have had to be investigated by mathematical modelling, because it would be very difficult to train people to walk with unusual values of q. Some other aspects of walking are more accessible to direct experiment. A man walking at a given speed could take

short strides at a high frequency, or longer strides at a lower frequency. Nevertheless, men of the same height walking at the same
speed generally use about the same stride frequency. Figure 3-5(b)
shows results from an experiment in which the rate of oxygen consumption of a man was measured. He walked on a conveyor belt
moving at constant speed, matching his speed to that of the belt
so that he remained stationary relative to the laboratory. He took
steps in time with a metronome which was run at several different frequencies so that the graph of energy consumption against
stride frequency could be drawn. The graph shows a minimum at
a stride frequency of 1.0 Hz. This is about the frequency which
the man used naturally, when walking at the same speed without
the metronome. This man (and the others who took part in the
experiment) used stride frequencies close to the optima which
minimized energy costs.

3.4 Gaits of dogs and sheep

Quadrupedal mammals make more than one change of gait, in
their range of speeds. Many of them change successively from a
walk to a trot to a canter and finally to a gallop. Only the change
from walk to trot, which corresponds to the human change from
walk to run, will be discussed.

People change gait very abruptly, making large changes both
of duty factor and of shape factor. The change of gait has accordingly been interpreted as a catastrophe, in the mathematical
sense (Alexander, 1989). Alexander and Jayes (1978) found that
when a dog changes from walking to trotting, the duty factor and
shape factor change less than in humans. When a sheep makes
the same change there is no sudden jump in either factor, but one
gait merges into the other. This suggests a speculation based on
catastrophe theory.

The catastrophe in humans is represented in Fig. 3-7(a). In this
graph, some index of gait is plotted against speed. (This index
might be shape factor or duty factor or some combination of the
two.) The line shows for each speed, what gaits give energy expenditure minimum or maximum values. This is the same kind of

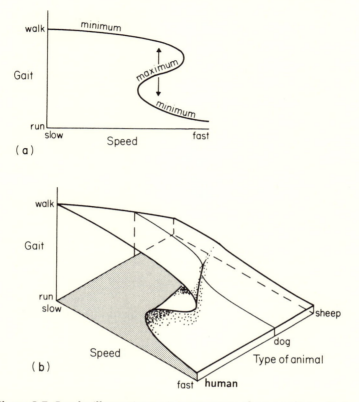

Figure 3-7. Graphs illustrating an interpretation of mammalian gaits in terms of catastrophe theory. They are explained in the text.

graph as Fig. 2-5(e) but has two folds instead of one. The optimum behaviour is to walk at low speeds and run at high ones, making the change abruptly where the folds overlap (at about the position of the double-headed arrow).

Figure 3-6(b) shows how dogs and sheep can be fitted into the scheme. An extra axis labelled 'type of animal' has been added, making the graph three-dimensional. The line has accordingly become a surface. It has been drawn with a pleat that peters out towards the back of the graph. If humans, dogs and sheep are represented by the positions marked on the 'type of animal' axis, humans should make a large abrupt gait change at the critical speed, dogs should make a smaller one and sheep should change gradually from a walk to a trot. This shows that there is no need to find a qualitative difference between dogs and sheep to explain the difference between an abrupt and a gradual gait change: a quantitative difference could suffice. So far, there is no indication as to what the difference might be or what properties of the animals are represented by the 'type of animal' axis. I cannot even predict how other animals should fit into the scheme. Catastrophe theory has done no more than suggest directions for further thought.

One of the complications that has been ignored in this discussion is that animals with short legs change gaits at lower speeds than ones with long legs. If the details of Fig. 3-7(b) were to be worked out more rigorously it would be necessary to represent speed by a dimensionless parameter that took account of differences of size.

Catastrophe theory makes a distinction between control variables and state variables. The optimum values for the state variables depend on the values of the control variables. Figure 2-5(e) has one variable of each type: the cost of failure F is the control variable and the factor of safety s is the state variable. In Fig. 3-7(b) there are two control variables (speed and type of animal) and one state variable (gait). Catastrophe theory says that when there is just one of each kind of variable, the only possible kind of catastrophe is the fold exemplified by Fig. 2-5(e). When there are two control variables and one state variable, the only possible kind of catastrophe is the cusp exemplified by Fig. 3-7(b). With more variables, other kinds of catastrophe are possible. Woodcock and Davis (1980) have written a very simple introduction to catastrophe theory and Saunders (1980) has written a more mathematical explanation.

3.5 Gaits for tortoises

The preceding sections of this chapter are about the energy cost of movement. This one is about equilibrium.

An animal standing still must be in equilibrium under the external forces that act on it, which are its weight and the forces on the soles of its feet. If the animal walks or runs at constant speed the resultant of the forces on its feet must have a mean value that is equal and opposite to its weight, but there is no need for perpetual equilibrium. A running animal is quite obviously not in equilibrium, at a stage of the stride when all its feet are off the ground.

Any departures from equilibrium during walking or running cause 'unwanted' displacements of the body, including fluctuations of height, pitching and rolling. (Pitching and rolling are explained by sketches beside Fig. 3-9.) At high speeds these displacements are too small to be troublesome, because a departure from equilibrium is quickly followed (later in the same stride) by a compensating departure in the opposite direction. Any unwanted component of velocity (or angular velocity) is soon reversed, and the amplitude of the unwanted displacements is kept small. At low speeds, strides last longer. Unwanted components of velocity remain uncorrected for longer and the displacements may become inconveniently large. An animal that walks very slowly must keep fairly near to equilibrium throughout the stride.

Tortoises walk exceedingly slowly, often taking two seconds for a stride. They pitch and roll quite noticeably, so they obviously do not maintain perpetual equilibrium. They have short legs and hold their shells quite close to the ground so they cannot pitch and roll very much, or allow their bodies to fall far, without hitting the ground. It is particularly necessary for them to avoid departing too far from equilibrium.

A tortoise can be in stable equilibrium with only three feet on the ground, but not with fewer (a three-legged stool is stable but a two-legged stool would not be). A four-legged animal like a tortoise could in theory move its feet one at a time so that it always had three on the ground and was always in stable equilibrium. It would have to move its feet in appropriate sequence so that the feet on the ground were always appropriately placed, relative to its centre of mass. It would move on a level course, neither pitching nor rolling.

This is not possible for real tortoises because it would require the forces on the feet to make large, instantaneous changes. Figure 3-8(a) shows the required pattern of forces. It is the only possible pattern, if the duty factor is 0.75. Some variation is possible at higher duty factors but the abrupt changes of force are still required. Tortoises have slow-acting muscles which cannot make abrupt changes of force. They have no need for fast muscles because they do not have to pursue prey or run from predators; they feed on plants and escape danger by withdrawing into their shells. Slow muscles are more economical of energy than fast ones, in maintaining tension. What is the best gait for an animal with slow muscles, which cannot walk as in Fig. 3-8(a)?

It is necessary to state the problem more precisely. Consider a quadruped whose feet exert only vertical forces. (Force platform records of tortoises walking show that the forces exerted by their feet keep much more nearly vertical than the forces on the feet of people. Jayes and Alexander, 1980.) Suppose that the muscles are so slow that the only pattern of force each foot can exert is the one shown in Fig. 3-4(a), with a shape factor of zero. This involves no abrupt changes of force. Suppose that the speed, stride length and duty factor have already been chosen. At what stage of the stride should each foot be moved?

The question requires an answer in terms of relative phases. The relative phase of a foot is the time during the stride at which it is set down, expressed as a fraction of the duration of the stride. The stride will be reckoned to start when the left fore foot is set down, so the relative phase of that foot will be, by definition, zero. What are the best relative phases for the other feet? Figure 3-8(a) shows (at the bottom) the relative phases that would make constant equilibrium possible, if the muscles were fast enough, but tortoises have very slow muscles.

The problem has still not been fully defined. How will we decide which relative phases are best? The problem is to minimize unwanted displacements, but three kinds of unwanted displacement have to be considered: pitch, roll and fluctuations of height. Let the angle of pitch (θ, Fig. 3-9) fluctuate in the course of a stride

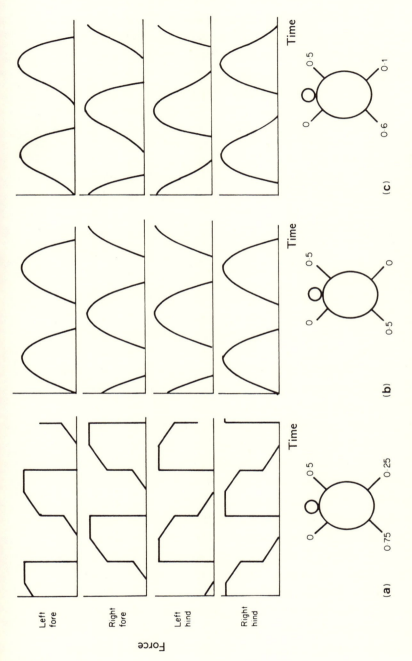

Figure 3-8. Graphs of vertical forces exerted by the feet of a quadruped, against time, for three different walking gaits. The diagrams below the graphs show the relative phases of the feet. The gaits are (a) the gait required for perpetual equilibrium, (b) the optimum gait for a quadruped with very slow muscles, as predicted by the mathematical model described in the text and (c) the gait used by tortoises.

between maximum and minimum values θ_{max} and θ_{min}. Let the angle of roll fluctuate between ϕ_{max} and ϕ_{min} and the height of the centre of mass between h_{max} and h_{min}. It seems likely that tortoises might choose the relative phases of their feet so as to minimize some function Φ

$$\Phi = F[(\theta_{max} - \theta_{min}), (\phi_{max} - \phi_{min}), (h_{max} - h_{min})] \qquad (3.7)$$

It is not at all obvious what form this function should take, except that an increase in any one of $(\theta_{max} - \theta_{min})$, $(\phi_{max} - \phi_{min})$ and $(h_{max} - h_{min})$ (leaving the others unchanged) should always increase Φ. Conveniently, we can be vague about this. It will be shown that there is no need to define the function more precisely.

Equations have been written that make it possible to calculate $(\theta_{max} - \theta_{min})$ etc. for any set of relative phases (Jayes and Alexander, 1980). Some results are shown in Fig. 3-8. they refer to gaits in which the relative phases are

left fore, 0   right fore, y
left hind, δ right hind, $(y + \delta)$

(i.e. the phase difference between the two hind feet is the same as between the fore feet). The contours on the graphs show that $y = \delta = 0.5$ is the only gait which keeps all three kinds of unwanted displacements small. Any likely form of the function Φ would have a minimum value for this gait. Not only is it the best of the whole range of gaits represented in Fig. 3-9: no better gait can be found by considering also gaits in which the fore and hind phase differences are different.

This optimum gait is shown in Fig. 3-8(b). Notice that diagonally opposite feet move simultaneously. It is quite different from the gait required for perpetual equilibrium (Fig. 3-8a), and more like the gait that tortoises actually use (Fig. 3-8c). However, the latter gait is a little different in that diagonally opposite feet do not move quite simultaneously. The discrepancy between the optimum predicted mathematically, and the actual gait, seems to be due to tortoise muscles being a little faster than was assumed.

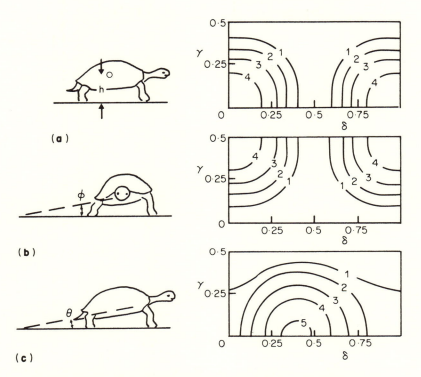

Figure 3-9. Graphs showing the ranges of unwanted displacements, for a quadruped walking as described in the text. The quantities γ and δ are relative phases. The contours show ranges in arbitrary units of (a) height of the centre of mass, h, (b) angle of roll, ϕ and (c) angle of pitch, θ. The diagrams on the left explain h, ϕ and θ. The duty factors of all the feet are 0.83. The graphs are from Jayes and Alexander (1980).

Force platform records of tortoises walking show that the force on each foot does not rise and fall symmetrically, as was assumed. Instead, graphs of force against time are skewed so as to be just a little more like the shape required for perpetual equilibrium. Further mathematics showed that with muscles just fast enough to do this, the optimum gait was very close indeed to the observed one.

This research leaves a further question unanswered. What is the optimum intrinsic speed for the leg muscles? Faster muscles would be less economical but would allow steadier walking and (if required) higher speeds.

4

Optimum behaviour

THE PREVIOUS CHAPTER was about the repetitive movements of walking and flight, but this one is about more complex behaviour. Should a bird eat only the biggest and best worms or should it eat every worm it finds? How long should a predator go on trying to get the remaining nourishment from one carcass before leaving it to look for other prey? Should a stag challenged by a rival fight or run away? These are just some of the questions asked in this chapter. Similar topics are discussed in books by Krebs & Davies (1993), Stephens & Krebs (1986), and Hughes (1993).

4.1 Choosing worms

Suppose a bird searching the ground for food has a choice of just two kinds of worm. Both are equally nutritious and equally easy to find, but one kind is bigger than the other, or easier to swallow. Should the bird eat this kind only, or should it eat both? It will be assumed that it is advantageous to make the rate of feeding (mass of worms per unit time) as large as possible. This problem was tackled by MacArthur and Pianka (1966) and subsequently by many other authors including Charnov (1976a).

Let unit area of ground have on it n_1 worms of type 1 (mass m_1 each) and n_2 worms of type 2 (mass m_2 each). If the bird never stopped to pick up a worm it could search this area in time T. If it feeds it needs additional time t_1 to pick up and swallow each type 1 worm, and t_2 for each type 2 worm.

Let m_1/t_1 be greater than m_2/t_2. This means that type 1 worms are bigger than type 2, or quicker to eat, or both. To maximize its rate of feeding, the bird must obviously eat every type 1 worm that it finds, but it is less obvious whether it should also eat type 2. Suppose it eats every type 1 worm and a fraction a of the type 2 worms. It spends time $(T + n_1t_1 + an_2t_2)$ searching unit area of ground, and eats a mass $(n_1m_1 + an_2m_2)$ of worms. Hence its rate of feeding, Q, is given by

$$Q = (n_1m_1 + an_2m_2)/(T + n_1t_1 + an_2t_2) \qquad (4.1)$$

What value of a gives the largest value of Q? Notice that a appears only in the final terms of the numerator and denominator. The rate Q is the same for all values of a if

$$m_2/t_2 = n_1m_1/(T + n_1t_1) \qquad (4.2)$$

Also, Q increases as a increases if

$$m_2/t_2 > n_1m_1/(T + n_1t_1) \qquad (4.3)$$

In this case the bird should give a the largest possible value, which is 1: it should eat all the type 2 worms. On the other hand, Q decreases as a increases if

$$m_2/t_2 < n_1m_1/(T + n_1t_1) \qquad (4.4)$$

and in this case the bird should give a the lowest possible value, which is zero: it should eat no type 2 worms.

Condition 4.3 can be re-arranged to give

$$m_2T + m_2n_1t_1 > n_1m_1t_2$$
$$n_1 < m_2T/(m_1t_2 - m_2t_1) \qquad (4.5)$$

If type 1 worms are sparse (n_1 small) the bird should eat all the type 2 worms, however many there are. If type 1 worms are plentiful it should eat none of type 2. The advantage changes abruptly at a particular value of n_1, so it is never advantageous to eat just some of the type 2 worms.

A similar conclusion applies if there are more than two kinds of worm. As the best kinds become sparser the bird should eat more and more of the less good kinds. In any particular case it should eat all of certain kinds, and none of the rest.

Do birds behave as the mathematical model suggests they should? In an experiment, caged great tits (*Parus major*) were offered a mixture of large and small pieces of mealworm, scattered on a moving belt (Krebs *et al.*, 1977). The small pieces had their handling time increased by attached tags of adhesive tape, which the birds had to remove before eating them. When the belt brought pieces at a low rate the tits were unselective: they ate large and small pieces in the same proportion as they were offered. When, however, there were plenty of large pieces the tits ate them almost exclusively, no matter how plentiful the small pieces were. The critical rate of delivery of large pieces, at which they made the change, was about the rate suggested by the theory.

That experiment shows that tits can behave as the mathematical model suggests, in a highly artificial laboratory situation. Do birds feeding in their natural habitats behave in the same way? The redshank (*Tringa totanus*) is a common European shore bird, which feeds largely on polychaete worms. Redshank feeding on mudflats were watched by telescope, and the sizes of the worms they ate were estimated by comparison with the length of the birds' bill (Goss-Custard, 1977). After several hours observation, samples of mud were taken so that the population densities of worms of different sizes could be determined.

The population densities and the relative abundance of large and small worms varied between shores. The redshank took longer to pick up and swallow large worms than smaller ones: for instance, 6 s for worms of dry mass 100 mg and 0.4 s for 2 mg worms. The ratio m/t (mass/handling time) was larger for the large worms, which should therefore have been eaten preferentially when sufficiently plentiful. It seemed that they were. Figure 4-1(a) shows that the number of large worms eaten, per metre walked by the redshank, was highly correlated with the population density of large worms. This is what would be expected if the redshank ate every large worm they found. The number of worms eaten per metre walked was always far less than the number per square metre of mud, because the strip of mud searched by a redshank walking in a straight line is much less than a metre

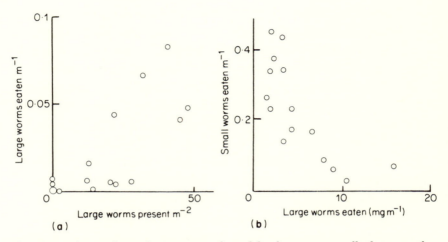

Figure 4-1. The numbers of worms eaten by redshank per metre walked over mud. (a) A graph of the number of large worms (>80 mg dry mass) eaten per metre, against the number of large worms per square metre of mud. (b) A graph of the number of small worms (0.5–10 mg dry mass) eaten per metre, against the dry mass of larger worms eaten per metre. Redrawn from Goss-Custard (1977).

wide and because many of the worms would be hidden, buried in the mud. The number of small worms eaten per metre walked was not significantly correlated with the population density of small worms but showed a strong negative correlation with the mass of larger worms eaten per metre (Fig. 4-1b). When plenty of large worms were available, the redshank ate rather few small ones. The mathematical model suggests that when large worms are plentiful, small ones should be entirely avoided, so the feeding behaviour of redshank seems to be a little different from the optimum.

4.2 Food for a moose

The worms which have been discussed differed in size and ease of swallowing but were assumed all to have the same composition. This section is about choosing between foods of different composition.

Moose (*Alces alces*) living in a National Park in Michigan feed in summer partly on terrestrial plants (mainly leaves of trees) and partly on aquatic plants. The latter have low energy content but they contain useful quantities of sodium, which is present only in trivial quantities in the terrestrial plants. Sodium is a necessary component of the diet, largely to replace sodium lost in the urine. Should moose eat only aquatic plants, or only terrestrial ones, or a mixture of the two? The following discussion is a simplified version of an analysis by Belovsky (1978).

Consider a typical adult moose of mass 358 kg. Males and females have different requirements and limitations but this discussion will be based on average values for adults of either sex. Let this animal eat every day A kg aquatic plants and B kg terrestrial plants. It has been shown by experiment that the energy a moose can extract from 1 kg food is about 0.8 MJ for aquatic plants (which contain a very large proportion of water) and 3.2 MJ for terrestrial plants. Hence the animal's daily intake of energy is $(0.8A + 3.2B)$ MJ. It will be assumed that the moose selects its diet so as to make this intake as large as possible.

In selecting its diet the moose must observe certain constraints. It has been calculated from sodium losses in the urine, requirements for growth etc. that a moose must eat on average at least 17 kg aquatic plants daily, to satisfy its need for sodium. It has also been calculated from measurements of the volume of the rumen and the time food takes to pass through it that the moose is incapable of digesting more than about 33 kg food daily. Thus the problem the moose has to solve, in choosing its diet, is

$$
\left.
\begin{array}{ll}
\text{maximize } (0.8A + 3.2B) & \text{(energy intake)} \\
\text{subject to } A \geqslant 17 & \text{(sodium requirement)} \\
\text{and } A + B \leqslant 33 & \text{(rumen capacity)}
\end{array}
\right\} \qquad (4.6)
$$

Problems like this can be solved by the mathematical technique of linear programming. This is particularly easy when, as in this case, only two variables have to be considered. The method is shown in Fig. 4-2. Any diet (i.e. any combination of values of A and B) is represented by a point on the graph. The thick lines

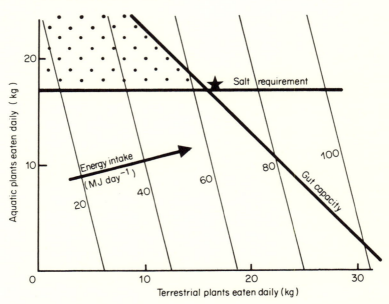

Figure 4-2. A diagram used to determine the optimum diet for a moose. The axes represent masses of aquatic plants (A kg) and terrestrial plants (B kg) eaten daily. The constraints represented by thick continuous lines restrict feasible diets to the stippled area. The broken lines are contours of daily energy intake (MJ day^{-1}). The star represents the actual diet of wild moose.

represent the constraints. The horizontal one is the line $A = 17$, so only points above it satisfy the moose's need for sodium. The thick diagonal line is $A + B = 33$, so only points below and to the left of it can be digested. Together, the two constraints limit feasible diets to the stippled area.

The parallel lines in Fig. 4-2 are contours of equal daily energy intake. They show that of all feasible diets, the one which gives maximum energy intake is represented by the bottom right-hand corner of the stippled area. This is the theoretical optimum diet, and consists of 17 kg aquatic plants and 16 kg terrestrial plants daily. The actual diet, determined by watching moose and surveying plants for evidence of cropping, is almost exactly the same as the theoretical one. It lies very slightly outside the stippled area, presumably due to an error in estimating either it or rumen capacity.

The original discussion of the moose's problem (Belovsky, 1978) was more complicated than the version presented here. Two kinds of terrestrial food were distinguished, herbs and the leaves of deciduous trees. Herbs cannot be eaten as quickly as the leaves of trees, but they yield a little more energy per unit mass. Also, account was taken of the danger of overheating in the midday sun. Moose seem to have to spend part of the day on land and part in water, to control body temperature in summer.

Since three kinds of food were being distinguished it was not possible to solve the problem by drawing a two-dimensional graph like Fig. 4-2. The equivalent graph was three-dimensional, with planes instead of lines representing the constraints. Such graphs are inconvenient to draw, but linear programming problems in three (or more) dimensions can be solved by algebraic methods (see, for instance, Koo, 1977). In two-dimensional problems the solution always occurs at the intersection of two of the constraint lines. In three-dimensional problems it occurs at the intersection of three constraint planes, and in multi-dimensional problems at an intersection of the hyperplanes that represent the constraints. The standard method of solution examines the intersections systematically, avoiding any that are obviously not the optimum.

The three-dimensional analysis indicated that in the optimum diet, about 90% of the terrestrial food should be leaves of trees and about 10% should be herbs. Moose were observed to eat these foods in almost exactly these proportions.

Belovsky (1978) also considered what the optimum diet would be if the aim were not to maximize energy intake, but to minimize time spent feeding. In that case the optimum diet would include no herbs.

4.3 When to give up

The larvae of ladybirds (coccinellid beetles) feed on aphids, eating out the soft tissues and leaving the hard exoskeleton. In the early stages feeding is quick and easy, but as the meal continues the remaining food becomes more difficult to extract. How long should a larva persist with one aphid before leaving it to look for another? It will be assumed as in section 4.1 that it is advantageous to make the mean rate of intake of food as large as possible. Larvae which feed fast will probably grow and develop faster than ones which feed slowly, and produce offspring sooner. Hence natural selection should favour behaviour which increases the rate of intake of food.

Figure 4-3(a) shows the evidence that during a meal, the remaining food becomes progressively more difficult to extract (Cook and Cockrell, 1978). Ladybird larvae which had previously been starved for 24 h were given an aphid each to feed on. Their meals were interrupted after various times, and the remains of the aphids weighed. Thus the graph for feeding on a single aphid was obtained. Notice that its gradient becomes smaller (the rate of extraction of food falls) as the meal proceeds. The other graph shows that this was not due to diminishing appetite: when fresh aphids were supplied at ten minute intervals, feeding continued at an almost constant rate for fifty minutes.

Let a ladybird larva spend time t feeding on each aphid, and let the mass of food it extracts in this time be $m(t)$ (The letter t in parentheses indicates that m is a function of t). Let the mean time required to find a new aphid, after leaving a partially-eaten one, be T. The mean rate of intake of food, Q is given by

$$Q = m(t)/(T + t) \qquad (4.7)$$

The optimum value of t is the one which makes Q a maximum, and could be found by calculus if we had an algebraic expression giving $m(t)$ as a function of t. Such an expression could be chosen to fit the graph in Fig. 4-3(a), that shows the time course of a typical meal on a single aphid.

Figure 4-3(b) shows a different, easier method of finding the optimum. The curve is a graph of $m(t)$ against t, copied from Fig. 4-3(a). Straight lines have been drawn from the point $(-T, 0)$ to intersect the curve at various values of t. The gradient of each

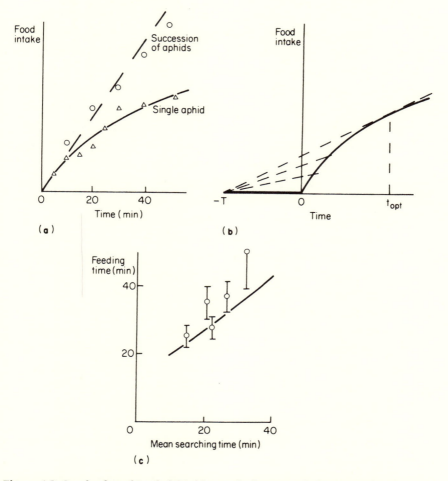

Figure 4-3. Graphs describing ladybird larvae feeding on aphids. (a) Graphs of the dry mass of food extracted against time. Graphs for a single meal and a series of ten minute meals on successive aphids are shown. (b) A diagram based on the graph for a single meal showing how the optimum duration of the meal (t_{opt}) can be determined. T is the time required to find an aphid. (c) A graph of feeding time t against search time T. The points show observed values and the line shows the theoretical optimum behaviour. The points in (a) and (c) have been taken from Cook & Cockrell (1978). In (c) they are shown ± one standard error.

of these lines is $m(t)/(T+t)$ so the steepest possible line, for any given value of T, indicates the optimum value of t. This line is the one which is a tangent to the curve.

Optimum values of t obtained in this way are shown in Fig. 4-3(c). These values are only rough estimates: different values would have been obtained if the smooth curve through the points in Fig. 4-3(a) had been drawn slightly differently.

Figure 4-3(c) also shows the mean feeding times t actually used by ladybird larvae. Between 2 and 32 aphids were distributed evenly on a tray, 0.5 m square. A ladybird larva was put on the tray and watched continuously for 4 h. Any aphid it ate was replaced immediately. The time spent feeding on each aphid and the time spent searching for the next one were recorded. Hence mean values of t and T were obtained. The more aphids there were on the tray, the shorter the searching time T and also the time t spent feeding on each aphid. The observed values of t are close to the graph of theoretical optimum values in Fig. 4-3(c). The ladybird larvae behaved approximately as the theory suggested they should.

A similar analysis has been applied to the mating of dungflies, *Sarcophaga* (Parker & Stuart, 1976). This is another situation in which persistence brings diminishing returns. Females arriving at cowpats to lay eggs are intercepted by males who copulate with them, although most of the females have sperm from previous matings in their spermathecae and could lay fertile eggs without further copulation. The longer the male prolongs copulation, the more of the sperm from previous matings are displaced and the larger the proportion of the offspring that are fathered by him. This proportion increases at a diminishing rate, just as the rate of feeding of a ladybird larva diminishes in the course of a meal. To maximize his number of offspring a male should copulate for a particular optimum time, and then go in search of another mate. It was estimated from field observations, using a graph like Fig 4-3(b) that the optimum duration for copulation in the particular circumstances was 41 minutes. The observed mean duration was only a little different: it was 36 minutes.

The theories for the ladybird larvae and dungflies are particular cases of the Marginal Value Theorem (Charnov, 1976b). This theorem refers to situations in which food or some other resource is patchily distributed, and exploitation of a patch gives diminishing returns. In the case of the ladybird larvae, the patches were individual aphids and the resource was their food content. In the case of the dungflies, the patches were individual females and the resource was unfertilized eggs. The theorem also assumes that when one patch is abandoned, appreciable time is needed to find or travel to the next. The theorem states that the rate of benefit is maximized by exploiting each patch until the rate of benefit from it falls to the maximum *mean* rate that can be sustained over a long period.

The theorem has also been applied to animals searching for hidden food. Cowie (1977) studied great tits (*Parus major*) looking for mealworms hidden in small jars of sawdust. The tits found the first few mealworms in each jar faster than they could find the remainder: the fewer the mealworms that remained, the longer on average it took to find each one. The tits moved on from each jar to the next at about the optimal times. The theorem could also be applied to natural patches of hidden food. For a bird feeding on caterpillars, each plant on which caterpillars might be found would be a separate patch.

The theorem applies only if patches are searched at random. If the prey are stationary and the predator searches each patch systematically, never going over the same ground twice, the number of prey removed has no effect on the rate of finding the remainder and the theorem does not apply.

4.4 Ideal free ducks

Imagine two people feeding the ducks in a pond. One, on the north bank, throws them bread at a rate (pieces per minute) r_N. The other, on the south, throws bread at a rate r_S. The ducks divide into two groups, n_N of them going to the north and n_S to the south. The ducks in the north group get bread at an average rate r_N/n_N and those in the south group at a rate r_S/n_S. If $r_N/n_N < r_S/n_S$, a duck in the north group can hope to improve its rate of feeding by moving to the south, and vice versa. Therefore, if each duck is trying to maximize its rate of feeding we should expect them to sort themselves out into groups such that $n_N/n_S = r_N/r_S$. This would be the "ideal free distribution."

The experiment has been tried (Harper, 1982). In one set of trials, bread pieces were thrown at 5-second intervals from both sides of the pond. It took about a minute for the groups of ducks to form and stabilize. From then on, the 33 ducks were about equally divided with (usually) 14–18 in each group. In other trials, bread was thrown twice as fast from one side as from the other and the ducks divided themselves into groups of about 22 and 11. In each case they kept very close to the ideal free distribution.

The analysis so far implies the assumption that the ducks are equal in competitive ability, so that the food in each patch is shared out equally among that patch's ducks. Now suppose what is likely to be true, that some of the ducks are dominant over others, able to intimidate subordinate ducks to give up food. In that case a duck may do best to avoid a group that includes dominant ducks, even though there may be fewer ducks in that group (Parker & Sutherland, 1986).

This effect has been observed. Harper (1982) identified six of the ducks as dominants and found that the other ducks were avoiding groups that included them, to an extent that would be explained if each dominant duck were regarded as equivalent to two non-dominants.

Bread being thrown into a pond is an unnatural situation, but analogous situations occur in nature. For example, in South American rivers, fruit-eating fish congregate under overhanging trees from which fruits are falling into the water. I do not know whether they adopt ideal free distributions: the question seems not to have been investigated. However, it has been shown that male dung flies, seeking mates, distribute themselves on cow pats and on the surrounding grass in an ideal free way (Parker, 1974b).

4.5 Two-armed bandits

Animals have to choose where to look for food. A redshank has to choose which part of the beach to feed on. A tit has to choose whether to hunt for insects on this tree or the next, and on which branch. How should animals choose between different feeding places when they do not know in advance how easily they will find food in them? The qualitative answer seems obvious enough: they should search for a while in each place, see how well they do in each and then concentrate on the place that seems best. The quantitative problem is more difficult.

How long should an animal spend trying the alternatives before making a decision? If it spends too little time trying them there is a serious danger of making a wrong decision on inadequate evidence: the animal may make the mistake of settling for the less good place. If it spends too much time trying the alternatives it is spending time on the less good place that could have been spent more profitably on the better one.

Problems like this would be very difficult to study in the complex situations that occur in nature. It seems better to start with simple experiments. Imagine a man offered the choice of two gambling machines (one-armed bandits). He tries both in an attempt to discover which is likely to give him the larger profit (or the smaller loss), and eventually settles for one of them. How and when should he make his choice? This is known as the two-armed bandit problem. It is a simple version of the problem faced by an animal choosing between feeding places.

Krebs, Kacelnik and Taylor (1978) devised an experiment to discover whether great tits can solve the two-armed bandit problem. They built a machine that dispensed small pieces of mealworm, when a bird hopped on a perch beside it. Each hop advanced the machine one step, presenting the bird either with a piece of mealworm or with an empty compartment. The machine was loaded with full and empty compartments in random order so that the bird could not predict the outcome of a particular hop. It could, however, assess the probability of being rewarded, from the results of previous hops.

Two of these machines were placed at opposite ends of an aviary and loaded to give rewards with different probabilities. A bird trained to use them was admitted to the aviary, and its hops on the two machines were recorded automatically on magnetic tape. It started as expected by trying both machines, making groups of a few hops on them alternately, but eventually settled for one machine and used only it. It nearly always chose the machine that gave the higher probability of reward. The experiment was performed with nine great tits, and five different pairs of probabilities of reward.

Figure 4-4 shows the mean numbers of hops taken before the bird made its decision. It was reckoned to have decided, when more than 90 of the next 100 hops were on the preferred machine. The probabilities for the two machines were always adjusted to add up to 0.5: if one machine gave rewards with probability P per hop, the other gave $(0.5 - P)$. Thus a difference in probability of 0.1, on the horizontal axis of the graph, implies that the probabilities for the two machines were 0.2 and 0.3. Similarly, a difference

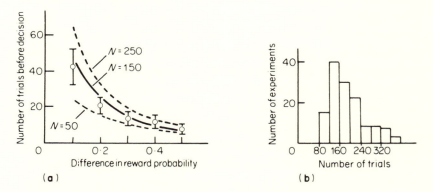

Figure 4-4. (a) A graph showing predictions and results for the two-armed bandit experiment. The number of trials made before the decision, is plotted against the difference between the values of P for the two machines. The lines show the optimum numbers of trials $(2n)$ for experiments with different total numbers of hops (N). The points are the geometric means of the numbers of trials made before decision, by nine great tits. The 95% confidence limits of the means are also shown. **(b)** A histogram showing the numbers of trials that the birds were allowed to make, in different experiments. Re-drawn from J. R. Krebs, A. Kacelink and P. Taylor (1978) *Nature, Lond.* **275**, 27–31. Reprinted by permission. © 1978 Macmillan Journals Limited.

Table 4.1. Rewards in a typical two-armed bandit experiment. In this example, machine A gives rewards with probability 0.2 per trial, and machine B with probability 0.3 per trial.

Trials per machine	Rewards from A	Rewards from B	Total reward so far from A	from B
1 to 5	10000	01101	1	3
6 to 10	00010	10000	2	4
11 to 15	10000	00010	3	5
16 to 20	00010	00010	4	6
21 to 25	01100	00001	6	7
26 to 30	10000	11000	7	9
31 to 35	00000	01010	7	11
36 to 40	00000	01100	7	13
41 to 45	10000	01001	8	15
46 to 50	00011	00100	10	16

of 0.5 implies values of 0 and 0.5. The smaller the difference, the more trial hops were taken.

Table 4-1 shows how the rewards might have been loaded in the machines, in a typical experiment. They have been arranged at random but with predetermined probabilities, by means of computer-generated random numbers. Suppose that the bird tried the machines alternately, taking bouts of five hops on each. After the first five hops on each it would have one reward from A and three from B and might decide to choose B. On the other hand it might feel that these few trials were inconclusive, and go on trying. If it continued and made 25 hops on each machine it would obtain six rewards from A and seven from B, and might well feel unable to make a confident decision even then. How long should it go on trying?

In this example, the total number of rewards from B stays consistently ahead of the number from A. The bird would be guided to the right decision even if it decided after only five trials on each machine. Suppose, however, that the sequence had started with trials number 11. In that case, the total reward from B would not overtake the total from A until 25 trials of each machine had been made. After five or ten trials of each machine the bird would have got equal numbers of rewards from each, and after 15 it would have got more from A. At what stage is it best to make the decision?

Since the machines give randomly-distributed rewards, the answer must involve probability theory. If the probability of a reward is P for each trial, the probability of no reward is $(1 - P)$ for each trial. The probability of getting exactly x rewards in n trials is

$$\Pi = P^x (1 - P)^{n-x} \binom{n}{x} \tag{4.8}$$

(The symbol $\binom{n}{x}$ means the number of possible combinations of choices of x items out of n). Apply this to the first five trials of machine A, which gave one reward out of five ($x = 1, n = 5$). If $P = 0.2$

$$\Pi = 0.2^1 (0.8)^4 \binom{5}{1} = 0.082 \binom{5}{1}$$

but if $P = 0.4$

$$\Pi = 0.4^1 (0.6)^4 \binom{5}{1} = 0.052 \binom{5}{1}$$

The probability of getting this result would be $0.082/0.052 = 1.6$ times as high if P were 0.2, as it would be if P were 0.4. Thus it might seem to a mathematically-minded bird that the probability of P being 0.2, was 1.6 times as high as the probability of P being 0.4. In this way, graphs can be produced showing the relative probabilities of different estimates of P, from the bird's point of view, after different numbers of trials.

Figure 4-5 consists of graphs like this. Initially before the bird has made any trials, all values of P between 0 and 1 are regarded as equally probable, for both machines. After five trials of each machine, with A giving 1 reward and B giving 3, the probability distributions have maxima at $1/5$ and $3/5$. At this stage the probability distributions have large standard deviations, but as trials continue their standard deviations diminish and P can be estimated more and more precisely. The bird can decide which machine is better, with increasing confidence.

The bird's best estimate of P, for a machine that has given x rewards in n trials, may be taken to be the mean of the probability distribution. It can be shown mathematically that this is $(x + 1)/(n + 2)$. The means of the probability distributions are indicated by broken lines in Fig. 4-5. They fluctuate quite a lot during the first few trials but eventually settle close to the true values of P.

Krebs and his colleagues (1978) suggested that the optimum behaviour for the bird would be as follows. It should try the two machines alternately and after each pair of trials estimate two total rewards:

(*i*) the total reward that would be obtained by using from then on, only the machine that at that stage seemed best and

(*ii*) the total reward that would be obtained by making one more trial of each machine, and then choosing the machine that seemed best. The bird should try the machines alternately until it reaches a stage at which estimate (*i*) exceeds estimate (*ii*). Thereafter it should use only the apparently better machine.

The estimates depend on the total number of hops N that the bird expects to make. After n trials of each machine, $(N - 2n)$ hops remain to be made. If the bird has had x_A rewards from machine A and x_B from machine B, the means of the probability distributions are $(x_A + 1)/(n + 2)$ and $(x_B + 1)/(n + 2)$ and estimate (*i*) is $(N - 2n)$ times the larger of these two means. Estimate (*ii*) requires a more complicated calculation that takes account of all the possible outcomes of the next two trials.

Krebs and his colleagues (1978) investigated this optimizing technique by computer simulation. They showed that the numbers of trials that should be made before the decision, were as shown by the lines in Fig. 4-4(a). Many more trials should be made before the decision if the machines gave very similar probabilities of reward (for instance, 0.2 and 0.3) than if they gave very different ones (for instance 0 and 0.5). Also, more trials should be

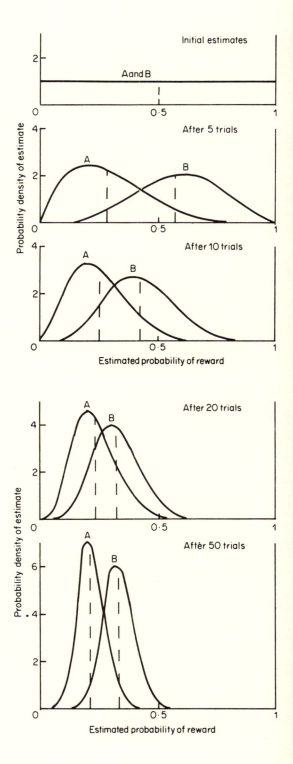

Figure 4-5. Graphs showing how a bird's estimates of the reward probabilities *P* of the two machines, might develop during the experiment described by Table 4-1. The probability densities of the estimates of *P* are plotted against the estimates, for five stages in the experiment.

made when N is large, than when it is small. The figure shows that the mean numbers of trials actually made by tits were close to the theoretical optimum values for $N = 150$. The number of hops allowed varied from 80 to about 350 in different experiments, with a mode of 150 (Fig. 4-4b). The tits behaved in the optimum manner for birds expecting to be allowed that number of hops.

In conventional statistics, a probability is an objective measure based solely on observed data. In Bayesian statistics, an initial estimate of probability is modified in the light of experience: the observer guesses the probability and then modifies his guess as he or she accumulates relevant data (Phillips, 1973). Krebs and his colleagues (1978) took a Bayesian attitude in their analysis of the two-armed bandit problem. For their main calculations, they assumed that the bird initially regarded all values of P between 0 and 1 as equally probable (Fig. 4-5a). This is an unlikely assumption as the values of P in the experiments the birds had previously experienced were always 0.5 or less. It was shown, however, that this would have little effect on the optimum number of hops provided the bird did not start with any very precise pre-conceived estimate of P.

This experiment showed the optimizing ability of tits, but in a highly artificial situation. A wild bird is unlikely to have only two possible feeding sites available: it is more likely to be faced by numerous patches of food. For instance, a tit in a wood has an enormous number of branches to choose from. In such circumstances it would not be sensible to sample all the patches before choosing one. It would be better to go from patch to patch, moving on quickly from ones found to contain little food but feeding longer at ones with plentiful food. If the patches can be searched systematically, the good ones should be searched completely before they are abandoned. If they have to be searched at random, and the rate of finding food falls as the patch is depleted, the marginal value theorem applies and the bird should try to judge the optimum time for moving on. Green (1980) discussed this sort of situation in a paper that combines elements of the marginal value and two-armed bandit approaches.

4.6 Hunger and thirst

A bird is both hungry and thirsty when, suddenly, it finds food and water. Should it drink until its thirst is quenched, and then eat? Should it eat until its appetite is satisfied, and then drink? Should it eat and drink by turns? The discussion that follows is based on a theory by Sibly and McFarland (1976). It uses a mathematical method that is described in section 1.6. If you skipped that section when you read chapter 1, you should read it now. (It is about the shortest path between two points.)

The bird has presumably evolved behaviour that tends to maximize its fitness. Starvation and thirst increase the probability of death (even if only by a little) and reduce fitness. Suppose the bird needs quantities x_1 and x_2 of food and water, respectively, to restore it to a normal, healthy condition. How much is its fitness reduced? The simplest plausible hypothesis would be that the reduction in fitness was proportional to $(x_1 + x_2)$, but a more general hypothesis will be adopted. It will be assumed that the reduction in fitness is given by

$$\Phi = x_1^a + bx_2^a \qquad (4.9)$$

where a and b are constants. If $a < 1$, a deficit of two units of food or water is less than twice as bad as a deficit of one unit, but if $a > 1$, a deficit of two units is more than twice as bad. The aim of the bird is to reduce Φ as fast as possible. The rate of change of Φ is, by differentiation

$$\frac{d\Phi}{dt} = ax_1^{a-1} \cdot \frac{dx_1}{dt} + abx_2^{a-1} \cdot \frac{dx_2}{dt} \qquad (4.10)$$

(Remember the rules of differentiation given by equations *1.6* and *1.12*.)

In this equation, $-dx_1/dt$ and $-dx_2/dt$ are the rates of eating and drinking. The bird can eat at a maximum rate r_1 or drink at

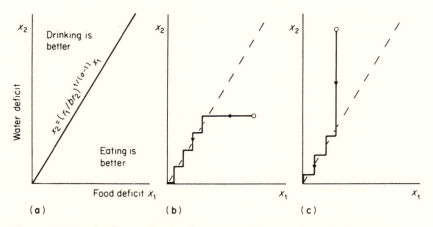

Figure 4-6. Graphs illustrating the discussion of hungry thirsty birds. Water deficit x_2 is plotted against food deficit x_1, in each case. (a) shows in which circumstances the bird should eat, and in which circumstances it should drink. (b) and (c) show the optimum path to satiation, from two different starting points.

a maximum rate r_2, but it cannot do both at once. It can reduce Φ at a rate $ar_1x_1^{a-1}$ by eating or at a rate $abr_2x_2^{a-1}$ by drinking (equation *4.10*). Eating is preferable to drinking if

$$r_1x_1^{a-1} > br_2x_2^{a-1} \qquad (4.11)$$

but drinking is preferable to eating if the reverse is true.

Figure 4-6(a) is a graph of x_2 against x_1, showing when eating is better and when drinking is better. It applies for any value of a greater than 1. It shows that a bird that is very hungry and only slightly thirsty should eat, but that one that is very thirsty and only slightly hungry should drink. If a were less than 1 the reverse would be true, which seems most unlikely. It will be assumed that a is greater than 1.

Figure 4-6(b) shows the optimal behaviour for a bird that starts from a point in the 'eating is better' region. It should eat, reducing x_1 until it just enters the 'drinking is better' region. Drinking a little will then bring it back into the 'eating is better' region and it should alternate between eating and drinking, zigzagging down the dividing line until its hunger and thirst are both satisfied. Similarly, the optimum behaviour for a bird starting in the 'drinking is better' region is to drink until it crosses the dividing line, and then zigzag down it (Fig. 4-6c).

Barbary doves (*Streptopelia*) behaved like this, in a series of experiments (Sibly, 1975). These experiments involved a highly artificial situation but nevertheless give some insight into dove behaviour. The doves were trained to use a machine (a Skinner box) that dispensed small quantities of grain and water. It dispensed grain when a red key was pecked and water when a green key was pecked. It could be adjusted to vary the maximum rates (r_1 and r_2) at which it would dispense grain and water.

Graphs very like Figs. 4-6(b) and (c) resulted from these experiments. The slopes of the zigzag lines in these graphs should be (according to the theory) $(r_1/br_2)^{1/(a-1)}$. It was found as predicted that the slope was increased when the machine was adjusted to increase r_1, and decreased when r_2 was increased. There was too much variation for it to be possible to calculate a accurately from the changes of slope, but the results suggested that a might be about 2 or 3.

When the green key was conveniently placed beside the red one, the bird could change back and forth between eating and drinking with little loss of time. The optimum behaviour at the boundary between the eating and drinking regions was to eat just a little and then drink just a little, moving down the boundary in small zigzags. If the keys were further apart more time would be lost in each change and it would be better to use bigger zigzags. Larkin and McFarland (1978) put partitions in the Skinner box so that the birds had to make detours of various lengths to get from one key to the other. The birds behaved appropriately. When the detours were longer, graphs like Fig. 4-6(b) and (c) showed larger zigzags.

It has been assumed so far that fitness depends only on the food and water deficits. Sibly and McFarland (1976) considered in addition the possibility that fitness might be reduced during feeding and drinking, especially when the bird ate or drank fast. This would be true in natural conditions, for instance, if a bird was less able to watch for predators while eating or drinking fast. In the experiments just described only slow feeding and drinking were permitted, and this complication was probably not impor- tant. Consider now a situation in which fast feeding is possible, but tends to reduce fitness. Feeding along will be considered with- out drinking, to simplify the analysis. Let the fitness of a hungry bird be reduced by an amount

$$\Phi = x_1^2 + (c.\dot{x}_1)^2 \qquad\qquad (4.12)$$

where x_1 is the food deficit as before, $\dot{x}_1$ is the rate at which it is changing (i.e. the rate of feeding is $-\dot{x}_1$) and c is a constant. The squaring of the two terms on the right hand side of the equation is a plausible guess which leads, as will be shown, to a realistic conclusion.

How fast should a hungry bird eat? This is a more difficult ques- tion than the previous one, because the bird has to balance the present disadvantage of fast eating against the future advantage of being less hungry. It has to find the optimum pathway from present hunger to future satisfaction. It will be shown that this is very like the problem discussed in section 1.6, of finding the shortest path between two points.

The longer a particular value of Φ persists, the more damaging it is. This suggests that the bird should aim to minimize the area under a graph of Φ against time. This area is

$$\Psi = \int_0^\infty \Phi.\mathrm{d}t \qquad\qquad (4.13)$$

Just as in the problem of the line between two points, the quantity to be minimized is an integral. The same method of solution can be used.

The solution must satisfy the Euler equation

$$\frac{\partial \Phi}{\partial x_1} - \frac{\mathrm{d}}{\mathrm{d}t}\left(\frac{\partial \Phi}{\partial \dot{x}_1}\right) = 0 \qquad\qquad (4.14)$$

Differentiation of Φ (equation 4.12) gives $\partial\Phi/\partial x_1 = 2x_1$ and

$\partial\Phi/\partial\dot{x}_1 = 2c^2\dot{x}_1$ so the Euler equation can be written

$$2x_1 - \frac{d}{dt}(2c^2\dot{x}_1) = 0$$

$$x_1 = \frac{d}{dt}(c^2\dot{x}_1) = c^2 \cdot \frac{d^2x_1}{dt^2} \qquad (4.15)$$

(remember that $\dot{x}_1$ means dx_1/dt). A method for solving equations like this is described in books on differential equations. There is a range of solutions, but only one solution is plausible because it is the only one that makes x_1 fall ultimately to zero. This solution is

$$x_1 = x_1(0)\exp(-t/c) \qquad (4.16)$$

where $x_1(0)$ means the value of x_1 at the beginning of the experiment. It is easy to check that this is a solution. By differentiating it twice we get

$$d^2x_1/dt^2 = (1/c^2)x_1(0)\exp(-t/c) = x_1/c^2 \qquad (4.17)$$

which is precisely the equation we set out to solve. It does not matter that there are other solutions. The optimum behaviour must satisfy the Euler equation but does not have to be the only behaviour that satisfies it.

Equation 4.16 shows that the optimum behaviour for a hungry bird with the opportunity to feed as fast as it likes, is to feed at a gradually decreasing rate, making x_1 fall exponentially towards zero.

4.7 Gamble when desperate

A bird that feeds only by day must accumulate enough food reserves before dusk to keep it alive through the night. Houston & McNamara (1988) discussed how it should behave, to minimize the risk of starvation. For small birds in winter, the danger may often be real.

Consider a very simple problem, even simpler than the one discussed by Houston & McNamara. To survive the night our imagi-
nary bird must have at least 10 units of energy reserves by 6 p.m.,
at which time it becomes too dark to feed. It has the choice of two
feeding places. At **A** the food supply is precisely predictable: the
bird will increase its energy reserves by one unit, for every hour
that it feeds. At **B**, however, it is unpredictable: in any particular
hour there is a 50% chance of adding two units to the energy reserves and a 50% chance of adding nothing. The average rate of
energy gain is the same at both places.

A bird that has 9 units of reserves at 5 p.m. (when one hour
remains for feeding) will certainly have the requisite 10 units by
6 p.m. if it feeds at **A**. If it feeds at **B**, it has only a 50% chance of
surviving. (It will either finish with 11 units and survive or with
9 units and die.) It should feed at **A**. However, a bird with only
8 units of reserves at 5 p.m. should feed at **B**, which gives it a 50%
chance of survival, whereas at **A** it would certainly die. It would
be in a desperate situation from which gambling offered the only
chance of escape. A bird with 7 or fewer units of reserves at 5 p.m.
is doomed, with no chance of survival.

The 5 p.m. column in Table 4-2 shows probabilities of surviving
the night for birds with different levels of energy reserves, that
spend the remaining hour feeding at **A** or at **B**. The bold type
emphasises (as explained above) that a bird with 9 units of energy
does better to feed at **A** and one with 8 units of energy at **B**.

Now consider the possibilities at 4 p.m. A bird with 9 units of
energy that feeds at A will have 10 units of energy already by
5 p.m. and will certainly survive. One that feeds at B will either
gain two more units of energy by 5 p.m. and survive; or it will gain
nothing in the first hour and still have only 9 units of energy at
5 p.m., in which case it can nevertheless guarantee its survival by
moving to A for the final hour.

A bird with 8 units of energy at 4 p.m. may feed at A for both
the remaining two hours, in which case it will finish with 10 units
and survive. If it feeds at B until 5 p.m. it will either have 10 units
by then and be certain of survival, or it will still have only 8 units,

Table 4.2. The optimum strategy for a bird that must accumulate 10 units of energy reserves by 6 p.m., if it is to survive the night. For each hour it must choose either the predictable strategy A or the unpredictable strategy B (details are given in the text). For each level of reserves at each time, the table shows the probabilities of survival for a bird that adopts strategies A or B for the next hour and thereafter uses the best strategy. Preferred strategies are indicated by bold type.

Reserves		2 p.m.	3 p.m.	4 p.m.	5 p.m.
10	A	1	1	1	1
	B	1	1	1	1
9	A	1	1	1	**1**
	B	1	1	1	0.5
8	A	1	1	**1**	0
	B	1	1	0.75	**0.5**
7	A	1	**1**	0.5	0
	B	1	0.75	0.5	0
6	A	**1**	0.5	0	0
	B	0.81	**0.63**	**0.25**	0
5	A	0.63	0.25	0	0
	B	0.63	0.25	0	0
4	A	0.25	0	0	0
	B	**0.38**	**0.13**	0	0

in which case its best option can be seen from the 5 p.m. column: remain at B and have a 0.5 chance of survival. Thus the mean probability of survival is 0.75, if the bird chooses B at 4 p.m. and then makes the best choice at 5 p.m.

The table has been built up in this way. As we go down each column we work out the probabilities of survival for a bird that for that hour chooses A or B, and makes whichever is the better choice for subsequent hours. Above the stepped line through the table the better choice (bold type) is A, and below it the better choice is B.

To work this out we had to work backwards through time, considering first the final hour, then the penultimate one, then the one before that. The reason for this is that for a bird with given energy reserves at a given time, the best strategy from then on is the same, however it reached that state. This technique for finding the optimum behaviour is called dynamic programming. It would of course be possible to work out the best strategy more precisely by using shorter steps of time, but the one-hour steps we have used are adequate to illustrate the principle.

Caraco, Martindale & Whittam (1980) performed experiments with juncos (*Junco*, a finch) to find out whether they were prepared to gamble when hungry. The birds had to choose between two covered dishes, one which they had been taught would always contain the same number of seeds, and the other which contained either twice that number or none, with equal probability. Juncos that had waited only an hour since their previous meal preferred the certainty, but ones that had been without food for four hours preferred to gamble.

4.8 Hunting lions

Dynamic programming has also been applied to the hunting be-haviour of lions (Mangel & Clark, 1988). They often fail to catch their prey, and thus risk starvation if they have a run of bad luck. Their food supply is more predictable if they hunt in groups. Ob-servations in the Serengeti have shown that a single lion hunting gazelle succeeds on only 15% of attempts, but the success rate for a pair of lions hunting together is 31%. That means that lions hunting in pairs can be more confident of a regular food supply, but against that must be set the disadvantage of having to share the kill.

Unlike small birds, lions can survive four or five days without food, without even having to call on their fat reserves. Their prob-lem is not one of surviving the night, but of long-term survival. Mangel & Clark (1988) used dynamic programming to calculate the probability of surviving 30 days, for lions that could choose each day whether to hunt singly, in pairs, or in larger groups. They concluded that when gazelles are the prey, a lion with a good deal

of food in its stomach from previous kills should choose the predictable option, hunting with a companion; but a lion with a near-
empty stomach should hunt alone. As in the case of the small
bird, the rule is: gamble when desperate. Hunting alone reduces
the chance of success, but if success comes the reward is greater.

Gazelles are small enough for one hungry lion to eat all the meat
off a carcass, but lions also take much larger prey such as zebra,
which are big enough to satisfy at least four hungry lions. Mangel
& Clark's (1988) model predicted that when hunting gazelles, lions
should never work in groups of more than two. Its predictions for
zebra depended on whether the hunters shared their prey with
other members of the pride.

Lions actually hunt in groups of very variable size. A long series of observations in the Serengeti showed that 51% of gazelle
hunts were undertaken by single lions, 22% by pairs, and 27% by
larger groups, and it is not known whether the lions that hunted
singly were the hungriest. However, the theory does help us to understand why group hunting occurs, and there is some evidence
from the Kalahari that groups break up when food is short.

4.9 Territories

Many animals establish territories and defend them against other members of the same species, and sometimes of other species
as well. A territory may be held by an individual, a mated pair
or a larger group. It may be used for feeding, mating, nesting or
several of these activities. It may be large or small. For example,
great tits hold territories of (typically) about one hectare but some
small fish living on coral reefs hold territories of only about one
square metre. What is the optimum size for a territory?

Here is a simple theory. Consider an animal that feeds exclusively on its own territory and excludes all competitors. Assume
that natural selection favours behaviour that maximizes the quantity of food collected daily. This seems a likely assumption for
nesting birds, which can rear more young if they collect more food,
or for birds preparing for migration, which can build up their en-

ergy reserves faster. Let the area of the territory be A and let the quantity of food available daily, per unit area, be p. Each day, the territory holder has a quantity pA of food available, if it has time to collect it all.

The territory may be defended by chasing intruders or by some warning display, such as the song of many birds. For instance, the great tit defends its territory by singing from various perches. The bigger the territory, the more time will be needed to defend it. Assume that the time required is simply proportional to the area, that it is kA where k is some constant. (It could be argued that it is more likely to be proportional to the perimeter than to the area, but we will make the simpler assumption.) Let the bird be active for time T each day (this will generally be the time from dawn to dusk) and let it collect food at a rate q when it is free to feed. If it spends time kA defending its territory, the time available for feeding is $(1 - kA)T$, and the quantity of food it has time to eat is $(1 - kA)qT$.

We have already seen that the quantity of food available is pA. Thus the quantity of food the bird can eat is pA or $(1 - kA)qT$, whichever is the smaller. Fig. 4-7 shows that daily food intake is greatest when

$$pA = (1 - kA)qT$$
$$A = qT/(p + kqT) \tag{4.18}$$

A territory of this size has just enough food to keep the bird busy throughout the day.

The theory could be made more elaborate and possibly more realistic; for example, by taking account of the energy cost of defending the territory or of the risk of injury. However, the very simple form will serve our purpose. Equation 4.18 predicts that if there is less food per unit area (if p is smaller) the optimum territory size will be larger.

The theory has been tested in experiments with hummingbirds (*Selasphorus rufus*), migrating through California (Carpenter, Paton & Hixon, 1983; Hixon, Carpenter & Paton, 1983). These birds break their journey for a week or two in the Sierra Nevada, where they establish temporary territories on alpine meadows and feed on the nectar of the Indian paintbrush (*Castilleja*), a plant that produces conspicuous clusters of bright red flowers. Each territory contains about 1000 to 4000 individual flowers (there are many flowers in each cluster). The birds spend an average of 22% of the day feeding from these flowers, 3% defending their territory, and 75% perching. The time spent perching seems to be needed for the nectar to be absorbed through the wall of the gut: the bird is not wasting time.

The rate at which the birds laid down energy reserves was monitored by providing instrumented perches, which automatically weighed the birds. Each bird varied the size of the territory it defended from day to day. When their territories were smaller than usual, the birds put on weight less fast. Three out of four birds also gained weight less fast when their territory was bigger than usual, than at intermediate territory sizes. These results are not as clear as might be wished, but give some indication that there is an optimum territory size that allows faster weight gain than would be possible if the territory were larger or smaller. This is what Fig. 4-7 led us to expect, but it should be remembered that only about 3% of the day is spent in territory defence. That suggests

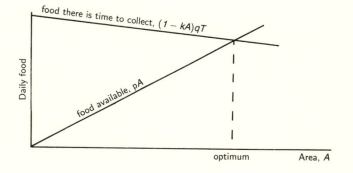

Figure 4-7. A schematic graph illustrating the theory of optimal territory size. The quantity of food available daily (pA) and the quantity that there is time to eat ($(1 - kA)qT$) are plotted against the area A.

that the disadvantage of having a bigger-than-optimal territory should be small.

Equation 4.18 led us to predict that if food were sparser (p were smaller), optimum territory size would be larger. This was investigated by covering half the flower heads in a territory with plastic bags so that the birds could not get at them. The birds responded as predicted, by enlarging their territories. Territory size was reduced again when the bags were removed. Control experiments in which non-flowering heads were bagged were used to check that the birds were responding to the reduced number of flowers, not merely to the presence of plastic bags.

4.10 Discretion or valour

Animals of the same species often come into conflict, for instance, when one bird enters the territory of another. Many of the animals involved have jaws, claws or horns capable of inflicting serious injuries, but dangerous fights are generally rare. For instance, a British robin (*Erithacus rubecula*) finding another intruding in its territory generally does not attack it but sings and adopts a posture that seems designed to display its red breast. The intruder generally then retires. Territorial disputes between adult robins are very seldom escalated to actual fights (Lack, 1946). Red deer stags disputing the ownership of harems do not at first fight, but exhibit themselves to each other adopting postures that seem adapted to showing off the size of their antlers (Maynard Smith and Price, 1973; Clutton-Brock, *et al.*, 1979). The animal with the smaller antlers generally retires at this stage without a fight, but if he does not, the stags interlock antlers and join in a pushing match which decides the issue. This is quite dangerous and injuries occur, but the animals do not seem to be trying to wound each other. Also, the branched antlers are far less dangerous weapons than sharp, unbranched ones would be. Why do animals not fight more readily and more dangerously?

People used to answer this question by saying that too much fighting would threaten the survival of the species. A species that

Table 4.3.

In a contest against:		Hawk	Dove
These animals obtain the scores shown	Hawk	$\frac{1}{2}(V - W)$	V
	Dove	0	$\frac{1}{2}V$

fought a lot would suffer a lot of injuries, so natural selection would favour more peaceful species. There is a fallacy here. Suppose a fighting mutation arose in a generally peaceful species. If the fighters were more successful than the pacifists in obtaining food and mates, and consequently produced more offspring, they would be favoured by natural selection. The proportion of fighters in the population would increase even though the increased frequency of injuries might lead eventually to extinction of the species.

In previous sections of this chapter it was possible to identify optimum behaviours which should be favoured by natural selection because they are in some sense better than the alternatives. This is not necessarily the case in contests: the best way to behave depends on how the opponent behaves. The same situation arises in games such as poker. Maynard Smith and Price (1973) and Maynard Smith (1974, 1982) showed how the mathematical theory of games could be adapted to treat the evolution of animal conflict behaviour.

Imagine a contest in which only two strategies are possible: *Hawk* and *Dove*. A Hawk always escalates a contest to an all-out fight. A Dove displays but never fights, and retreats if attacked. Which strategy will be favoured by natural selection?

Since the theory of games is to be used, a system of scoring is needed. Let the winner of a contest score V (victory). Let a loser who gives in without a fight score 0 but let one who is beaten in a fight score $-W$ (wound). When a Hawk meets a Dove the Hawk always wins (scoring V) and the Dove retires unhurt (scoring 0). When two Hawks meet, one scores V and the other $-W$

so the mean score for a Hawk against a Hawk is $\frac{1}{2}(V - W)$. When two doves meet the contest is decided by some means other than fighting. One scores V and the other 0 so the mean score is $\frac{1}{2}V$. These payoffs are shown in Table 4-3. If the reward for winning exceeds the penalty for losing a fight (i.e. if $V > W$), Hawk is the better strategy both against Hawks and against Doves, and will be favoured by natural selection. If $V < W$, however, Hawk is the better strategy against Dove but Dove is the better strategy against Hawk.

A new concept had to be added to the theory of games to deal with the evolution of conflict behaviour. This is the concept of the evolutionarily stable strategy (ESS). A strategy is an ESS if in a population of animals that adopted it, no mutant strategy would be favoured by natural selection. In a population of Hawks with $V < W$, Dove mutants would be favoured, so Hawk is not an ESS. In a population of Doves, Hawks would be favoured, so Dove is not an ESS. A mixed population would tend to evolve until it contained a fraction V/W of Hawks and a fraction $(W - V)/W$ of Doves. In this mixture, Hawks and Doves would obtain equal mean scores.

Other strategies might arise by mutation. One of the possibilities is the strategy that Maynard Smith (1979) called Bourgeois. This depends on the assumption that in every contest, one animal can be identified unambiguously as the owner of the disputed resource, and the other as an interloper. For instance, a territorial bird might count as owner in his own territory and interloper in any other bird's territory. The Bourgeois strategy is: play Hawk when you are the owner, but Dove when you are the interloper. A Bourgeois fights to defend his own possessions, but not to rob others of their possessions.

Table 4-4 shows the payoffs in contests between Hawks, Doves and Bourgeois. In contests with Hawks or Doves, a Bourgeois has probabilities 0.5 of being the owner and 0.5 of being the interloper. The probabilities that he will behave as a Hawk or as a Dove are each 0.5, so his expected score is the mean of the scores he would get by behaving consistently as a Hawk, or as a Dove. Similarly, in contests against Bourgeois, Hawks and Doves get mean scores

Table 4.4.

In a contest against:		Hawk	Dove	Bourgeois
These animals obtain the score shown	Hawk	$\frac{1}{2}(V-W)$	V	$\frac{1}{2}V + \frac{1}{4}(V-W)$
	Dove	0	$\frac{1}{2}V$	$\frac{1}{4}V$
	Bourgeois	$\frac{1}{4}(V-W)$	$\frac{3}{4}V$	$\frac{1}{2}V$

which are the means of the scores they would get against Hawks and against Doves. In a contest between two Bourgeois, the owner scores V and the interloper retires unhurt, scoring 0. The mean score for Bourgeois against Bourgeois is therefore $\frac{1}{2}V$. The table shows that if $V < W$, Hawk is the least favoured strategy in a population of Hawks, so hawk is not an ESS. Dove is the least favoured in a population of Doves, so Dove is not an ESS. Bourgeois, however, is the most favoured strategy in a population of Bourgeois, so it is an ESS.

In a fight between the owner of a territory and an interloper, the owner might have an advantage. He might know the ground better than his opponent, or he might have had the opportunity to occupy the best defensive position. Notice that no such assumption was made in compiling Table 4-4. Bourgeois is an ESS even if it is assumed, as here, that ownership gives no advantage at all in a fight. Indeed, any obvious asymmetry between the competitors could be the basis of an ESS. All that is necessary is that it should be possible in any contest to identify one animal as the 'A' animal and the other as the 'B' animal. The strategy 'if A, play Hawk but if B, play Dove' is an ESS against Hawk and Dove. The converse strategy 'if A, play Dove but if B, play Hawk' is also an ESS, and whichever of the two becomes established first will continue indefinitely.

This can be shown by referring again to Table 4-4. Suppose that in a population of Bourgeois, a mutant Anarchist arises. This mutant behaves in opposite fashion to Bourgeois: it plays Dove when it is the owner and Hawk when it is the interloper. Because it is initially rare, nearly all its contests are with Bourgeois. In half of them it would be the interloper and play Hawk, and the Bourgeois owner would also play Hawk. In half it would be the owner and play Dove, and the Bourgeois interloper would also play Dove. Its score would be the mean of Hawk against Hawk and Dove against Dove, $(\frac{1}{2}V - \frac{1}{4}W)$. This is less than the mean score of Bourgeois in a population of Bourgeois, so Anarchist would be eliminated by natural selection.

This argument could be extended to suggest that Anarchist is as likely to evolve as Bourgeois and that whichever is established first will survive. This may be misleading. It has already been suggested that owners are likely to have an advantage in fights. Suppose Bourgeois and Anarchist mutants both appear before either has become predominant in the population. Bourgeois will always fight as an owner and will therefore probably win more than half its fights. Anarchist will always fight as an interloper and will probably lose more than half its fights. Consequently Bourgeois will be preferred by natural selection. Another possibility is that the reward for winning, V, may be larger for an owner than for an interloper (Parker, 1974a). For instance, the owner of a territory may have spent time and energy familiarizing himself with the territory and building a nest. He would have no need to repeat these activities after winning a contest but an interloper would have to start from scratch. If V is larger for owners than for interlopers, Bourgeois is more likely to evolve than Anarchist. Robins and many other territorial birds behave very much like Bourgeois. Interlopers meeting territory owners generally retreat without fighting. This may be an example of Bourgeois strategy, but it may not: the interloper may not retreat because he has adopted the Bourgeois strategy, but because he has lost previous fights with the same owner and expects that he would lose if he fought again.

One of the most convincing demonstrations of Bourgeois behaviour concerns speckled wood butterflies (*Pararge aegeria*), which live in woods in Britain (Davies, 1978). Males spend the night among the leaves of the trees, but on sunny days many of them fly down to ground level and perch in patches of sunlight. Females wander around the wood, preferentially visiting patches of sun-

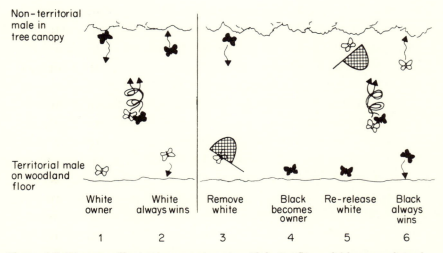

Figure 4-8. Diagrams illustrating experiments with butterflies, which are explained in the text. In these diagrams, one individual butterfly is shown black and another white. In the actual experiments, the butterflies were distinguished by different coloured spots of paint on their wings. From Davies (1978).

light. Males are more likely to have an opportunity to mate if they wait in a patch of sunlight, than if they wait elsewhere. A male will not tolerate another in the same patch unless the patch is very large (more than about 15 m²). As time passes and the patch moves over the floor of the wood, the owner male moves with it.

The butterflies in a wood were marked with spots of paint on their wings so that they could be recognized individually. There were more males than patches of sunlight: at any instant, about 60% of the males were on the ground in patches of sunlight, and 40% were higher among the leaves of the trees. From time to time the latter flew down to investigate patches of sunlight. If they found the patch unoccupied they stayed there, but if they found a male already there the two males flew together for a few seconds in a spiral flight (Fig. 4-8). The owner always won this brief contest: he retained the patch and the interloper flew away.

On ten occasions a male was caught and removed from a patch, and another allowed to take his place. Ten seconds after the new owner had landed, the original one was replaced. There was a spiral flight of about the usual duration and the new owner retained the patch. This happened even when the new owner was a male who had approached as an intruder when the original owner was in possession, and lost that contest.

These observations suggest Bourgeois behaviour but do not exclude the possibility that current ownership, however recently acquired, makes it easier to win a contest. An even stronger indication that the Bourgeois convention was being used came from another (more difficult) experiment, performed on five occasions. A second male was introduced surreptitiously into an already-occupied patch and made to perch there, unaware of and unnoticed by the original owner. This procedure was intended to give each male the impression that he himself was the owner. If both adopted the Bourgeois strategy, neither should be prepared to concede victory to the other. This seemed to be the case. When the males eventually noticed each other they made an exceptionally long spiral flight (mean duration 40 s instead of the usual 4 s).

Butterflies seem ill-equipped for fighting and may not be in much danger of being wounded, even in a long contest. The penalty corresponding to W in Tables 4-3 and 4-4 may not be physical damage, but loss of time and energy. Both contestants lose time and energy in a long contest, but this would not prevent Bourgeois from being an ESS.

The Bourgeois strategy involves sudden changes of behaviour whenever a frontier is crossed. Colgan, Nowell and Stokes (1981) have shown that these changes can be regarded as mathematical catastrophes and have formulated a catastrophe theory model of territorial fighting.

Yet another strategy might arise if there was a reliable way of predicting the outcome of a fight. This strategy is Assessor: fight if you expect to win but not otherwise. Suppose that Assessor arises as a mutation in a population of Bourgeois. The payoffs in this situation are shown in Table 4-5. In contests between two Bourgeois or between two Assessors there are no fights because both con-

Table 4.5.

In a contest against:		Bourgeois	Assessor
These animals obtain the scores shown	Bourgeois	$\frac{1}{2}V$	$\frac{3}{8}V - \frac{1}{4}W$
	Assessor	$\frac{5}{8}V$	$\frac{1}{2}V$

Table 4.6. Calculations for the Bourgeois/Assessor game.

Case	Owner	Stronger animal	Bourgeois score	Assessor score
a	Bourgeois	Bourgeois	V	0
b	Assessor	Assessor	0	V
c	Bourgeois	Assessor	$-W$	V
d	Assessor	Bourgeois	$\frac{1}{2}V$	$\frac{1}{2}V$
mean			$\frac{3}{8}V - \frac{1}{4}W$	$\frac{5}{8}V$

testants adopt the same convention. One scores V and the other 0 so the mean score is $\frac{1}{2}V$. For contests between Bourgeois and Assessor there are four possibilities which will be assumed to be equally probable (Table 4-6). In case (a) the Bourgeois is both the owner and the stronger animal so he wins by common consent. He scores V and the Assessor scores 0. In case (b) the Assessor wins by common consent. In case (c) the Bourgeois is the owner but the Assessor is the stronger animal so both are prepared to fight. The Assessor, being stronger, wins. He scores V and the Bourgeois scores $-W$. In case (d) neither animal is prepared to fight and the contest is decided at random, without injury. Each animal scores V or 0 with equal probability. The means of the scores obtained in these four cases are the scores shown in Table 4-5. This Table shows that if $V < W$, Assessor does better than Bourgeois, both against Bourgeois and against other Assessors. Natural selection should favour Assessor, and Bourgeois should be eliminated from the population. It was assumed in the table that the outcome of fights could be predicted with certainty but Assessor is still favoured when prediction is imperfect (see Maynard Smith and Parker, 1976).

The Assessor strategy seems to be used by various animals, for instance by stags which seem to use antler size as an indicator of fighting ability. Similarly, wild sheep (*Ovis dalli*) seem to use horn size (Geist, 1966). Groups of sheep seem to have a social hierarchy determined by horn size and all copulations are performed by the older, larger-horned rams. Strange rams have been observed joining existing groups and were ignored or accepted both by larger-horned and by smaller-horned rams. Nearly all the fights that occurred were between opponents of very similar horn size.

5

Optimum life-styles

THIS CHAPTER asks questions about reproduction. Is it better to lay a lot of small eggs or a few large ones? How much of the available food energy should be used for growth, at each stage of life, and how much for reproduction? What are the best proportions of male and female offspring? Is it better to look after your children or to abandon them and breed again? Sibly & Calow (1986), Stearns (1992), and Charnov (1993) discuss many of the issues raised in this chapter.

The chapter makes repeated use of a mathematical method which is an extension of methods described in sections 1.5 and 1.6. If you skipped these sections when you read chapter 1, you should read them now. (One is about the shapes of tin cans and the other about the shortest path between two points.)

5.1 How many eggs?

Suppose that a female animal has a limited amount of material for producing eggs. Should she lay a lot of small eggs or a few large ones? The small eggs would give her more offspring, if they all survived, but the large ones might have a better chance of survival. Pianka (1976) discussed this question.

The answer depends on the relationship between the size of an egg and the probability of its producing a mature adult. Three possible relationships are shown in Fig. 5-1. In each case the probability is shown as zero below a certain size because there must be a minimum size below which survival is impossible. (At the very least, the egg must be large enough to contain the necessary genetic material.) For eggs above the minimum size it is assumed that larger eggs always have higher probability of survival. Apart from this, it is not immediately obvious what shape the graphs should be.

Suppose an animal uses a mass m of material to produce n eggs of mass m/n each. If each egg has probability P of producing a mature adult, the fitness of the parent is Pn. Consider a point $(m/n, P)$ on the graph in Fig. 5-1(a). The gradient of a straight line through this point and the origin is Pn/m. If m is constant this gradient is proportional to the parent's fitness and animals can be expected to evolve to make the gradient as large as possible.

In Fig. 5-1(a) the graph becomes progressively steeper, so that the larger the egg, the greater Pn/m. To maximize its fitness the animal should use all the available material in a single large egg. Many animals lay clutches of more than one egg, and this graph is presumably not realistic for them. In Figs. 5-1 (b and c) the graphs curve the other way, giving Pn/m a maximum value at a moderate egg size ($n > 1$). The maximum is indicated by the tangent through the origin, just as the optimum time to give up in Fig. 4-3(b) was indicated by a tangent. In Fig. 5-1(b), large eggs survive much better than small ones and the optimum egg size is large. In Fig. 5-1(c), quite small eggs survive almost as well as large ones and the optimum size is smaller.

Two examples will be used to test the hypothesis implied by Fig. 5-1(b) and (c). In both cases the animals give birth to larvae instead of laying eggs, but this does not affect the analysis.

Guppies (*Poecilia reticulata*) are tiny freshwater fish, often kept in home aquaria. Two groups of wild populations living in Trinidad were preyed on by different predatory fish; in one case *Crenicichla* (which eats mainly adult guppies) and in the other *Rivulus* (which prefers juveniles). Guppies from *Rivulus* streams gave birth to fewer, larger young than those from *Crenicichla* streams, which is what one would expect because large young survive much better than small ones in these streams. To confirm that the difference in offspring size and number could be explained by the different predators, Reznick, Bryga & Endler (1990) moved 200 guppies from a *Crenicichla* stream to a *Rivulus* one where there had previously been no guppies. They flourished until, 11 years (and 30–60

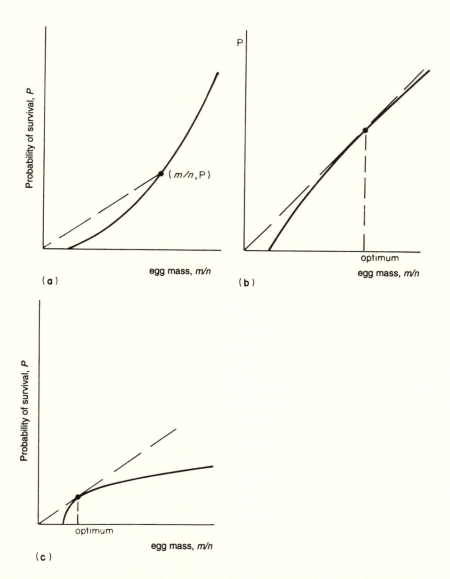

Figure 5-1. Schematic graphs showing possible relationships between the mass of an egg and the probability that it will survive to produce a mature adult. Egg mass is represented by m/n, where m is the total mass of the clutch and n the number of eggs in the clutch.

generations) later, samples were collected from both the original stream and from the new one. These were bred in identical conditions in the laboratory for two generations (to ensure that any differences between them were hereditary), and then their offspring were compared. It was found that the population that had been moved to the *Rivulus* stream now produced larger, fewer young than the population in the *Crenicichla* stream. Natural selection had apparently acted as predicted by the theory.

Our second example is a group of flies. The typical sarcophagids are the flesh flies (*Sarcophaga* etc.) which deposit their larvae on dung, carrion or open wounds. These sources of food occur unpredictably and do not remain available for long, but larvae deposited on them generally have a huge supply of potential food on which they can grow rapidly. It seems likely that the relationship between size at birth and probability of survival resembles Fig. 5-1(c). A fly which finds a good site will generally produce more adult offspring if it deposits a lot of small larvae, than if it deposits a few large ones.

A North American sarcophagid (*Blaesoxipha fletcheri*) has a very different way of life. Pitcher plants (*Sarracenia*) have vase-shaped leaves ('pitchers') which get filled by rainwater. Nectar secreted at the mouths of the pitchers attracts ants and other insects which often slip into the water, are unable to climb out, and drown. They are digested by enzymes secreted by the plant, which absorbs the products of digestion. The fly *Blaesoxipha* deposits its larvae in the pitchers. These larvae are not harmed by the enzymes but thrive in the pitchers, feeding on such other insects as are trapped. They live just below the surface of the water, breathing air through spiracles at their posterior ends.

In Ontario, *Blaesoxipha* larvae are most plentiful in June (Forsyth and Robertson, 1975). In this month, 85% of *Sarracenia purpurea* pitchers contained larvae. Nearly all of these (93%) contained just one larva. In a series of experiments, two larvae each were put into pitchers or plastic cups of water. When larvae met they fought, coiling round each other and apparently trying to submerge each other's spiracles. Within twelve hours, one larva nearly always climbed out of the container or was killed by drowning. The remaining, victorious larva was nearly always the larger of the two.

Blaesoxipha larvae which are large at birth are presumably much more likely to survive than small larvae. A large larva is unlikely to be overcome by another larva subsequently deposited in the same pitcher. Also, a large larva deposited in a pitcher which already contains a small one may be able to capture the pitcher. Since few pitchers remain vacant, competition for pitchers is probably intense. It seems likely that a graph of probability of survival against size at birth would resemble Fig. 5-1(b).

These arguments suggest that to maximize fitness, *Sarcophaga* should produce numerous small larvae and *Blaesoxipha* a few large ones. They do this. Three species of *Sarcophaga* each produce mean numbers of larvae between 50 and 84. *Blaesoxipha* produces a mean of only 11 larvae. The newborn larvae of various species of *Sarcophaga* have mean lengths ranging from 2.1 to 4.6 mm. *Blaesoxipha* adults are smaller than most *Sarcophaga* adults but their larvae are much larger, which a mean length of 7.0 mm (Forsyth and Robertson, 1975).

5.2 When to make queens

This section is about the annual cycle of reproduction of social wasps (Vespinae). Queens are the only wasps that survive winter, in the temperate zone. They are reproductive females, fertilized in autumn by males whose sperm they store. They spend the winter in a torpid state and establish new colonies in spring. Each colony is established by a single queen. She builds a small nest, lays eggs in it, and rears a small first brood of young, bringing them the food they need. Thereafter she continues to lay eggs, but most of the work of enlarging the nest, collecting food and caring for young is done by workers. These are females which do not reproduce.

All the members of the colony, except the queen, are her offspring. She has three kinds of offspring, which are reproductive females (next year's potential queens), reproductive males and workers. The reproductives are larger than the workers, and are reared in larger cells in the nest. They do not share in the work of the colony but are entirely responsible for transmitting its genes to subsequent generations.

The more reproductives a colony produces this year, the more colonies are likely to carry its genes next year. The fitness of colonies is probably, therefore, about proportional to the number of reproductives produced. Reproductives can be produced very much faster if the queen is assisted by a large team of workers, than if she does all the nest-building and food-collecting herself. There is therefore a great advantage in producing workers as well as reproductives. If fitness is to be maximized, what proportion of the offspring should be workers, and what proportion should be reproductives, at each stage of the season?

Macevitz and Oster (1976) attempted to answer the question by devising a mathematical model. The simplest form of their model is presented here. It depends on the following assumptions.

(i) The rate of producing offspring is not limited by the rate at which the queen can lay eggs, but depends only on the rate at which the workers can collect food.

(ii) The rate at which each worker can collect food is constant throughout the season, irrespective of the number of workers.

(iii) The mortality rate of workers is constant throughout the season, and the mortality rate of reproductives is small enough to be ignored. (Workers are liable to be killed by predators when they are out foraging, but reproductives seldom leave the safety of the nest.)

At time t, let the colony consist of n workers and N reproductives. At the beginning of the season, $t = 0$, $n = 1$ and $N = 0$. (The queen is counted as a worker because she undertakes, at this stage, all the tasks which workers perform later on.) At the end of the season, $t = T$, and $N = N_T$. The fitness of the colony is maximized by making N_T as large as possible.

At time t there are n workers. If all the food collected by them (in excess of their own requirements) were used for producing new workers, these would be produced at a rate rn. Alternatively, if all the food were used for producing reproductives, these would be produced at a rate Rn. Let a fraction u of the food be devoted to producing workers, and $(1 - u)$ to producing reproductives, so that the rates of producing workers and reproductives are urn and $(1 - u)Rn$, respectively. Workers die at a rate μn.

The rate of production of reproductives is

$$dN/dt = (1 - u)Rn$$

so the number produced in a short interval from time t to time $(t + \delta t)$ is $(1 - u)Rn.\delta t$, and the number at the end of the season is

$$N_T = \int_0^T (1 - u)Rn.dt \qquad (5.1)$$

The rate of increase in the number of workers is the difference between the rate at which they are being produced and the rate at which they are dying

$$dn/dt = (ur - \mu)n. \qquad (5.2)$$

Thus the problem is to choose u throughout the season, so as to

$$\left.\begin{array}{l} \text{maximize } N_T = \displaystyle\int_0^T (1 - u)Rn.dt \\ \text{subject to } dn/dt = (ur - \mu)n \end{array}\right\} \qquad (5.3)$$

This problem resembles the problem of the shortest path (section 1.6), in that the function to be maximized or minimized is an integral. It differs from that problem and resembles the problem of the food cans (section 1.5) in being subject to a constraint. The classical calculus of variations was used to solve the problem of the shortest path, but cannot solve this problem because of the constraint. The simple method of Lagrange multipliers was used for the problem of the cans but cannot be used here because of the integral. The method to be used combines elements of both these other methods. It is called Pontryagin's method, and deals with problems of the following type.

The state of a system at time t is represented by some quantity n (the number of workers, in our problem). The behaviour of the system is controlled by a quantity u that can be varied. It is required to adjust the time course of u so as to maximize the integral, between specified times, of some function $f(n, u, t)$ of n, u and t. There is an equation showing how the rate of change of state of the system, dn/dt, is affected by the control u. This equation (the constraint) gives dn/dt as a function $g(n, u, t)$. Thus a typical problem is

$$\left. \begin{array}{l} \text{maximize } \Phi = \displaystyle\int_{t_1}^{t_2} f(n, u, t).dt \\ \text{subject to } dn/dt = g(n, u, t) \end{array} \right\} \qquad (5.4)$$

(This is a simple example. More complicated problems can also be solved.) The Pontryagin maximum principle says that if u is made to vary with time in the optimum way, a function called the Hamiltonian is maximized at all times. This function is

$$\left. \begin{array}{c} H = f(n, u, t) + \lambda(t).g(n, u, t) \\ \text{where } \lambda(t) \text{ is a factor that changes at a rate} \\ d\lambda(t)/dt = -\partial H/\partial n \\ \text{and is zero at time } t_2 \\ \lambda(t_2) = 0 \end{array} \right\} \qquad (5.5)$$

The factor $\lambda(t)$ does the same sort of job as a Lagrange multiplier, but it is a function that varies with time, while a Lagrange multiplier has a constant value.

The Hamiltonian for the problem of equations 5.3 is

$$H = (1 - u)Rn + \lambda(t)(ur - \mu)n \qquad (5.6)$$

To find the optimum breeding strategy we need to find how u must be varied, so as to maximize H throughout the season.

The Pontryagin method is fully explained in books on optimization such as Koo (1977). However, it may be helpful to give here a

rough indication of how it works. In the problem we are considering, the aim is to produce as many reproductives as possible; but
it is also helpful to produce as many workers as possible, so long
as there is time for them to rear reproductives. In equation 5.6, the
term $(1 - u)Rn$ is the rate at which reproductives are being produced. The term $(ur - \mu)n$ is the rate of increase of the number
of workers. We want to make both these terms large. Early on, it
is very useful to increase the number of workers but at the end of
the season there is no point in producing workers because there
will not be time for them to rear young. For this reason, we use
$\lambda(t)$ to indicate the relative benefits of producing workers and reproductives. Early in the season workers are most valuable ($\lambda(t)$
is large) but at the end of the season producing them is pointless
($\lambda(t) = 0$).

Now we return to the problem of maximizing the Hamiltonian
H. Equation 5.6 tells us that a graph of H against u (with the
other quantities kept constant) would be a straight line. This implies that H is always maximized either by giving u the largest
possible value (if $\partial H/\partial u$ is positive) or the smallest possible value
(if $\partial H/\partial u$ is negative). At every stage throughout the season the
colony should devote the available resources either entirely to producing workers ($u = 1$) or entirely to producing reproductives
($u = 0$).

Differentiation of equation 5.6 shows that $\partial H/\partial u$ is
$[-R + \lambda(t).r]n$. This must be negative late in the season, when
$\lambda(t)$ is small and approaching zero. (Equations 5.5 show that $\lambda(t)$
is zero at the end of the season.) Thus u should be made zero
in the late part of the season. Earlier in the season $\lambda(t)$ may be
large enough to make $\partial H/\partial u$ positive, when u should be given the
value 1. The optimum strategy is to produce workers only ($u = 1$)
until some critical time τ, and thereafter to produce reproductives
only ($u = 0$). Solutions like this, which require the control variable to be held at one extreme and then switched suddenly to the
other, are called bang-bang controls.

Now that we know that a bang-bang control is required, we can
find the optimum switching time τ by simple calculus. From $t = 0$

to $t = \tau$, $u = 1$ so equation 5.2 gives

$$dn/dt = (r - \mu)n \quad (0 \leqslant t \leqslant \tau)$$

The solution of this differential equation is

$$n = \exp[(r - \mu)t] \quad (0 \leqslant t \leqslant \tau) \qquad (5.7)$$

(Remember that $n = 1$ when $t = 0$). Thus the number of workers increases exponentially: if for instance it doubles from 50 to 100 in one month, it doubles again to 200 in the next. The number at time τ is

$$n_\tau = \exp[(r - \mu)\tau]$$

After time τ, $u = 0$ and equation 5.2 gives

$$dn/dt = -\mu n \quad (\tau \leqslant t \leqslant T)$$

The solution of this equation that makes $n = n_\tau$ when $t = \tau$ is

$$n = n_\tau \exp[-\mu(t - \tau)]$$

By substituting for n_τ we get

$$\begin{aligned} n &= \exp[(r - \mu)\tau]\exp[-\mu(t - \tau)] \\ &= \exp[(r - \mu)\tau - \mu(t - \tau)] \\ &= \exp(r\tau - \mu t) \quad (\tau \leqslant t \leqslant T) \qquad (5.8) \end{aligned}$$

The number of workers declines exponentially, as workers die and are not replaced. The number of reproductives produced in the season can now be calculated from equation 5.1. Since $u = 1$ up to time τ, and 0 thereafter, this equation can be written

$$\begin{aligned} N_T &= \int_\tau^T Rn.dt \\ &= \int_\tau^T R.\exp(r\tau - \mu t).dt \end{aligned}$$

(using the value of n from equation 5.8). Integration gives

$$N_T = (R/\mu)\{\exp[(r - \mu)\tau] - \exp(r\tau - \mu T)\} \qquad (5.9)$$

The number of reproductives produced (N_T) can be maximized by choosing a particular switching time (τ) at which $dN_T/d\tau = 0$. By differentiating equation 5.9

$$
\begin{aligned}
dN_T/d\tau &= (R/\mu)\{(r - \mu)\exp[(r - \mu)\tau] - r\exp(r\tau - \mu T)\} \\
&= (R/\mu)\exp(r\tau)\{(r - \mu)\exp(-\mu\tau) - r\exp(-\mu T)\}
\end{aligned}
$$

This shows that $dN_T/d\tau = 0$ when

$$
\begin{aligned}
(r - \mu)\exp(-\mu\tau) &= r\exp(-\mu T) \\
(r - \mu)/r &= \exp(\mu\tau - \mu T) \\
\log_e[(r - \mu)/r] &= \mu(\tau - T) \\
\tau &= T + (1/\mu)\log_e[(r - \mu)/r] \qquad (5.10)
\end{aligned}
$$

This is the switching time that maximizes N_T. The number of reproductives produced will be greatest if workers only are produced up to this time, and reproductives only thereafter.

As the theory predicts, social wasps in general produce workers only until late in the season, when they produce reproductives. The predictions of the theory have been checked quantitatively in the case of a hornet, *Vespa orientalis*. Ishay, Bytinski-Salz and Shulov (1967) studied colonies of hornets nesting in boxes with transparent floors. They counted the adult hornets from time to time and found that their numbers changed during the year as shown in Fig. 5-2(a). No reproductives (other than the original queen) appeared until October. Between May 15 and September 15 the number of workers increased about exponentially from about 10 to 250. This implies that $(r - \mu)$ is 0.027 day^{-1} (see equation 5.7). Young hornets were marked with paint and their presence in the colonies was checked from time to time. Half had disappeared (presumably died) after 30 days. This implies that $\mu = 0.023$ day^{-1} ($= -\log_e(0.5/30)$), making r 0.050 day^{-1}. With

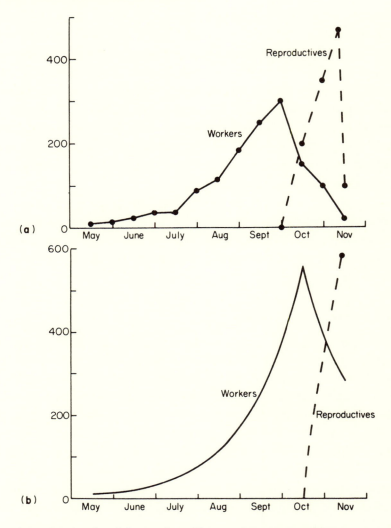

Figure 5-2. Graphs of the number of hornets (*Vespa orientalis*) in a colony, against time. Workers and reproductives are shown separately. (a) shows observed mean numbers from Ishay, *et al.* (1967) and (b) shows numbers in a colony adopting the theoretical optimum strategy as described in the text.

these values of r and μ, equation 5.10 gives an optimum $(T - \tau)$ of 27 days: reproductives should be produced only in the last 27 days of egg production. Fig. 5-2(a) shows that adult reproductives appeared over a period of about 40 days. This period cannot be measured accurately from the data and may be an over-estimate for any one colony, if the colonies whose numbers were averaged were out of synchrony, but it does seem likely that production of reproductives started slightly earlier than the date that the theory suggests as optimal. Fig. 5-2(b) shows how numbers should have grown in a colony behaving optimally. It is based on the values of r and μ already calculated together with $R = r$ (This is probably an over-estimate: reproductives are larger than workers and presumably cost more to produce).

Though the numbers of *Vespa* workers increase about exponentially, as the crude model requires, those of another wasp (*Polistes*) do not. Macevicz and Oster (1976) showed that a more complicated model was needed to simulate the history of *Polistes* colonies.

5.3 Growing or breeding

The previous section discussed how the resources of a colony of wasps should be divided, between producing workers and producing reproductives. This section discusses a related question: how should an individual animal divide its resources, between growth and reproduction, to maximize its fitness? Should it grow without reproducing until it reaches a certain size, and thereafter reproduce without growing? Alternatively, should it grow and reproduce simultaneously for some or all of its life? This problem is more complicated than the problem of the wasps because individuals live for variable times, whereas wasp colonies survive for precisely one season.

It seems necessary to think more about fitness, before tackling the problem. Imagine a population of animals in an environment where food and space are so plentiful that they do not restrict population growth at all. (This was assumed for wasps in assumption (*ii*), p. 110.) If the population has n_0 members at time 0 and n members at time t, the rate of increase dn/dt is proportional to n,

$$dn/dt = rn \qquad (5.11)$$

where r is a constant. This implies that the population increases exponentially

$$n = n_0 \exp(rt) \qquad (5.12)$$

This cannot go on forever. The population increase must lead eventually to overcrowding or a shortage of food. It must slow down and finally stop. Ecologists often take account of this by modifying equation *5.11*, writing

$$dn/dt = rn(K - n)/K \qquad (5.13)$$

where K is a constant. If n is much smaller than K, $K - n \simeq K$ and the equation is almost identical with *5.11*. As n increases, dn/dt first increases and then decreases again. When $n = K$, $dn/dt = 0$, so K is the largest population that the habitat can support. The solution of equation *5.13* is

$$n = \frac{K}{1 + [(K/n_0) - 1]\exp(-rt)} \qquad (5.14)$$

The population growth predicted by this equation is shown in Fig. 5-3(a). Growth very like this has been demonstrated in laboratory cultures of various organisms including protozoans (e.g. *Paramecium*) and beetles (e.g. *Tribolium*) (Krebs, 1985).

Figure 5-3(b) and (c) shows fluctuations of numbers in populations of two hypothetical species of animal. The species represented in (b) lives in a very stable habitat, and its numbers fluctuate only a little about the equilibrium value K. The species represented in (c) occupies temporary habitats that remain suitable only for a short time. When a habitat becomes suitable it is colonized by a few animals, which breed there. Their numbers increase exponentially, but never approach equilibrium because the habitat becomes unsuitable again. The animals can survive there no longer and most of them die, but some may escape and find an alternative habitat in a suitable state.

Many large species behave more or less as in Fig. 5-3(b). In an extreme case, the population of albatrosses (*Diomedea exulans*)

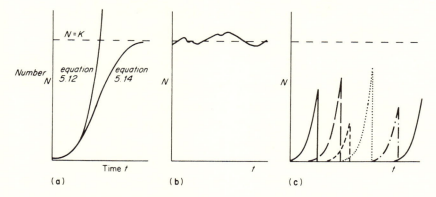

Figure 5-3. (a) Graphs of the number N of animals in a population, against time t, according to equations *5.12* and *5.14*. (b) and (c) Schematic graphs of number against time for two populations of animals.

on Gough Island (South Atlantic) is believed to have remained almost constant at 4000, from 1889 at least until 1971. The pitcher-plant fly (section 5.1) is also like this, in that the number of larvae approximates to the number that the habitat can support. Many other small species behave much more like Fig. 5-3(c). The flesh flies that spend their larval life in corpses or piles of dung are obvious examples.

Equations *5.11* to *5.14* were formulated for complete populations, but in any real population there would be differences between genotypes which would be acted on by natural selection. Some of the differences might affect maximum rates of reproduction and could be described by assigning different values of r to different genotypes. Others might affect ability to survive with limited resources, and could be described by assigning different values of K. In the situation represented by Fig. 5-3(b), fitness would be affected little by different values of r but considerably by different values of K: natural selection would favour genotypes with high K. Populations like this are described as K-selected (MacArthur and Wilson, 1967). In the situation represented by Fig. 5-3(c) fitness might depend largely on the ability to find a new habitat when the current one deteriorated. On the other hand, the finding of new habitats might be largely random, unaffected by genotype, and in this case fitness would depend mainly on r: the proportion of the faster-multiplying genotype would gradually increase. Species like this are described as r-selected.

The theories presented in this and the next section show how differences of life history strategy may give different genotypes different values of r, but take no account of possible differences in K (see Stearns, 1992, on this point).

We are ready now to proceed with this section's main question: how should a species divide its resources between growing and breeding? What follows is based on a paper by León (1976).

Consider a population in which females are being born, at time t, at a rate $v(t)$. Thus $v(t).\delta t$ females are born between times t and $(t + \delta t)$. Let the probability that a female will survive to age x be $P(x)$ and let the rate at which females of this age give birth be $b(x)$. The aim of this section is to find the optimum relationship between $b(x)$ and x.

Females born at time $(t - x)$ have age x at time t. The number of females born between times $(t - x)$ and $(t - x + \delta x)$ was $v(t - x).\delta x$. At time t, $P(x).v(t - x).\delta x$ of them survive and are giving birth to females at a rate $P(x).v(t - x).b(x).\delta x$. Animals being born at time t may in principle have mothers of any age between 0 and ∞, so the rate of birth of females is

$$v(t) = \int_0^\infty P(x).v(t - x).b(x).\mathrm{d}x \qquad (5.15)$$

If the population is increasing exponentially

$$v(t) = v(t - x).\exp(rx)$$

and division of both sides of equation 5.15 by $v(t)$ gives

$$1 = \int_0^\infty P(x).\exp(-rx).b(x).\mathrm{d}x \qquad (5.16)$$

Natural selection tends to maximize r, but unfortunately the equation cannot be solved for r: it cannot be written in the form

$r = \cdots$. The optimum strategy for reproduction can nevertheless be discovered, by the following procedure. Find the function $b(x)$ that maximizes the function

$$\Phi = \int_0^\infty P(x).\exp(-Rx).b(x).dx \qquad (5.17)$$

where R is an arbitrary constant. Do this for a range of different values of R and find the one for which the maximum value of Φ is 1. The number R and the reproductive strategy $b(x)$ obtained in this way, are the maximum attainable value of r and the corresponding strategy.

It is necessary to make some assumption about $P(x)$. The simplest plausible assumption is the one made for worker wasps in section 5.2, that there is a constant mortality rate μ. This implies

$$dP(x)/dx = -\mu P(x)$$
$$P(x) = \exp(-\mu x)$$

(This makes $P(0) = 1$ as required by the definition of $P(x)$.) Thus equation *5.17* becomes

$$\Phi = \int_0^\infty \exp[-x(R + \mu)].b(x).dx \qquad (5.18)$$

It is also necessary to be specific about the range of strategies $b(x)$ that are possible. An animal needs a certain food intake merely to maintain itself, but any excess can be used for growth or reproduction. Let an animal of mass m obtain food energy in excess of maintenance requirements at a rate $E(m)$. This notation implies that the rate is some function of m; it is presumably larger for larger animals, but is not necessarily proportional to m. Let animals of age x devote a fraction $u(x)$ of their excess energy to growth and a fraction $[1 - u(x)]$ to reproduction. They will presumably grow at a rate proportional to the rate of energy investment

$$dm/dx = C.u(x).E(m) \qquad (5.19)$$

where C is a constant. The rate of reproduction $b(x)$ may be proportional to the rate of energy investment in reproduction $[1 - u(x)]E(m)$, or it may be some other function of the investment: it does not necessarily require exactly twice as much energy to bear and rear twins as to bear and rear a single child. We will start with the simple assumption that it is proportional

$$b(x) = k[1 - u(x)]E(m) \qquad (5.20)$$

where k is a constant.

The optimization problem can now be formulated more precisely. Choose a function $u(x)$ so as to

$$\left.\begin{array}{l} \text{maximize } \Phi = \displaystyle\int_0^\infty \exp[-x(R + \mu)].k[1 - u(x)].E(m)\,\mathrm{d}x \\ \text{subject to } \mathrm{d}m/\mathrm{d}x = C.u(x).E(m) \end{array}\right\} \qquad (5.21)$$

At any age x, $u(x)$ may have any value between 0 and 1. This problem, like the problem of the wasps (section 5.2) can be tackled by Pontryagin's method. The state variable is m and the control variable is $u(x)$. The Hamiltonian is

$$\begin{aligned} H = {}& \exp[-x(R + \mu)].k[1 - u(x)].E(m) \\ & + \lambda(x).C.u(x).E(m) \end{aligned} \qquad (5.22)$$

Here, $\exp[-x(R + \mu)].k.[1 - u(x)].E(m)$ is the rate at which females of age x are producing young, and $C.u(x).E(m)$ is the rate at which they are growing. Growing now enables females to produce more young later, so we would like both these terms to be large; however, the advantage of growing diminishes to zero at the end of the animal's life. This change in the relative values of reproduction and growth is reflected by the falling value of the weighting factor $\lambda(x)$.

Equation 5.22 looks complicated but leads to a very simple solution. As for equation 5.6, a graph of H against $u(x)$ (with the other quantities kept constant) would be a straight line. This im-

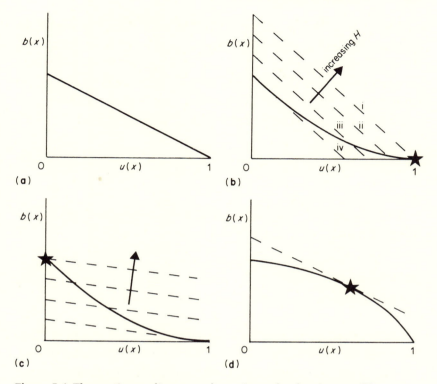

Figure 5-4. The continuous lines are schematic graphs showing possible relationships between the rate of giving birth $b(x)$ and the fraction of available energy that is devoted to growth $u(x)$. The broken lines are explained in the text. The star indicates the optimum strategy in each case.

plies that the optimal control is bang-bang. The animals should devote all available energy to growth until they reach a certain size. Thereafter they should grow no more, but devote all available energy to reproduction.

This result depends on the assumption of equation 5.20, which implies that a graph of $b(x)$ against $u(x)$ (for a given value of $E(m)$) is a straight line (Fig. 5-4a). Two other possibilities are shown in Fig. 5-4(b) and (d). Graph (b) implies that two offspring are less than twice as expensive as one and graph (d) that two are more than twice as expensive than one.

If we want to make assumptions like these, we must write equation 5.22 in a more general form

$$H = \exp[-x(R + \mu)].b(x) + \lambda(x).C.u(x).E(m)$$

where $b(x)$, the rate of reproduction, may be any function of age x. By re-arranging this we get

$$b(x) = [H - \lambda(x).C.u(x).E(m)]\exp[x(R + \mu)] \qquad (5.23)$$

This is an equation of a straight-line graph of $b(x)$ against $u(x)$, with slope proportional to $-\lambda(x)$ and intercept proportional to H. Several such graphs are represented by broken lines in Fig. 5-4(b). Each of these lines connects points having the same Hamiltonian. The Hamiltonian for any point on the continuous curve is the one for the broken line that intersects that point.

All the broken lines in Fig. 5-4(b) intersect the continuous line at some point between the no-growth end $(u(x) = 0)$ and the no-reproduction end $(u(x) = 1)$. They all have the same value of $\lambda(x)$ but different values of H. Line (i) has the largest intercept and therefore the highest value of H, and cuts the continuous line at the no-reproduction end. If the gradient (and so $\lambda(x)$) has been chosen correctly, the optimum strategy for the animal at this stage in its life history is to devote all available energy to growth.

Figure 5-4 (c) is identical with (b), except that $\lambda(x)$ is smaller and the broken lines slope less steeply. In this case the broken line that has the highest intercept, among those that cut the continuous line, cuts it at the no-growth end. This implies that to maximize the Hamiltonian, the animal should devote all available energy to reproduction.

If a graph of $b(x)$ against $u(x)$ curves as in Fig. 5-4(b), (c) (i.e. if d^2b/du^2 is always positive), the Hamiltonian is always maximized either by choosing $u(x) = 0$ or by choosing $u(x) = 1$: the optimal control is bang-bang, as for the straight line graph. This is not the case if the graph curves as in Fig. 5-4(d) (d^2b/du^2 negative). In this case it is possible for intermediate behaviour involving simultaneous growth and reproduction to be optimal. The broken line in this figure has a larger intercept than any line parallel to it, intersecting the continuous curve, would have.

The conclusion from this analysis is that for an r-selected species in the circumstances described, an optimal reproductive strategy can only involve simultaneous growth and reproduction if Fig. 5-4(d) applies: that is, if doubling the rate of producing offspring more than doubles the energy cost of reproduction. Otherwise, the optimal control is bang-bang.

If the energy required for reproduction consisted solely of food energy incorporated in eggs or fed to young, it would presumably be proportional to the number of offspring. A graph of $b(x)$ against $u(x)$ would be a straight line like Fig. 5-4(a). Some animals use additional energy in reproduction that is not proportional to the number of offspring being reared simultaneously. For instance many animals build nests or excavate burrows for their offspring. Little more energy may be needed to make a nest for a large clutch of eggs, than for a small one. Consequently, economies of scale are likely. Large families will be cheaper (per head) than small ones, and graphs of $b(x)$ against $u(x)$ will curve like the continuous lines in Fig. 5-4(b and c). It is difficult to think of circumstances that would make large families more expensive of energy (per head) than small ones, and give graphs like Fig. 5-4(d). (Remember that one of the assumptions of the model implies that energy intake is fixed: only the partition of energy between growth and reproduction is varied.) Thus it is difficult to think of circumstances within the limitations of the model that would not favour bang-bang control.

Many animals cease growth when they start reproducing. They include the insects, birds and mammals. In contrast, most fish continue growth throughout life. Some of the toothcarps (Cyprinodontidae) are exceptional, ceasing growth when they reach adult size (Brown, 1957).

5.4 Breeding and survival

How realistic is the model that has just been discussed? Interaction between reproduction and growth has been considered, but

may not reproduction affect mortality? May not the effort of reproduction increase the danger of death? The example of Pacific
salmon (*Onchorhynchus*), which die after reproducing for the first
and only time, suggests the possibility. The idea for the following
discussion came from Shaffer (1974) but the mathematical approach is quite different from his. Yet another approach is taken
by Bell (1980).

Consider the possibility that reproduction increases mortality.
To keep the discussion reasonably simple it will be assumed at this
stage that reproduction has no effect on growth. Thus an animal
that produces offspring rapidly will be supposed to grow as fast as
one that does not, but to be likely to die younger. To grow as fast
as a non-reproducing animal, it presumably collects more food.

As before, the problem is to maximize the function Φ of equation *5.17*

$$\Phi = \int_0^\infty P(x) \exp(-Rx) b(x) dx$$

Let the maximum possible rate of producing offspring, at age x,
be $B(x)$. This presumably increases with age, as the animal grows.
Let the actual rate of producing offspring be a fraction $w(x)$ of
this rate, so that

$$b(x) = w(x)B(x) \qquad (0 \leqslant w(x) \leqslant 1)$$

Instead of a constant mortality rate μ, assume that the mortality
rate is some function of $w(x)$: let it be $\mu(w)$. Then

$$dP(x)/dx = -P(x)\mu(w)$$

The optimization problem is thus

$$\left.\begin{array}{l} \text{maximize } \Phi = \int_0^\infty P(x) \exp(-Rx) w(x) B(x)\, dx \\ \text{subject to } dP(x)/dx = -P(x)\mu(w) \end{array}\right\} \qquad (5.24)$$

This is another problem that can be solved by Pontryagin's method.
The state variable is $P(x)$ and the control variable is $w(x)$. The
Hamiltonian is

$$H = P(x)[\exp(-Rx)w(x)B(x) - \lambda(x)\mu(w)] \qquad (5.25)$$

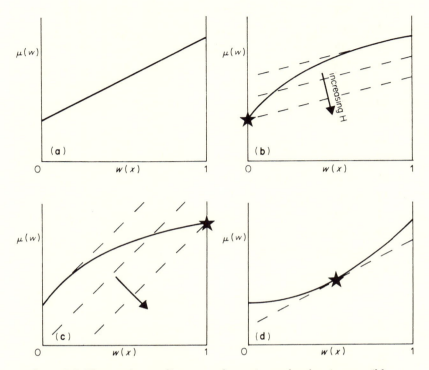

Figure 5-5. The continuous lines are schematic graphs showing possible relationships between mortality $\mu(w)$ and reproductive effort $w(x)$. The broken lines are explained in the text.

It has been assumed implicitly that $\mu(w)$ increases as $w(x)$ increases: faster reproduction implies higher mortality. Apart from this, it is not at all clear what the relationship between $\mu(w)$ and $w(x)$ should be like. If a graph of $\mu(w)$ against $w(x)$ was a straight line (Fig. 5-5a), a graph of H against $w(x)$ would also be a straight line and the optimal control would be bang-bang. Figure 5-5 also shows two other possible relationships. The optimal controls for them will be sought by the method already applied to Fig. 5-4.

Equation *5.27* can be written

$$\mu(w) = [1/\lambda(x)]\{\exp(-Rx)B(x)w(x) - [H/P(x)]\} \qquad (5.26)$$

which describes a graph of $\mu(w)$ against $w(x)$ with gradient proportional to $1/\lambda(x)$ and intercept proportional to $-H/\lambda(x)$. Such graphs are represented by broken lines in Fig. 5-5. They have all been drawn with positive gradients, implying that $\lambda(x)$ is positive, so that H is maximized by making the intercept $-H/\lambda(x)$ as *small* as possible. When a graph of $\mu(x)$ against $w(x)$ has the shape shown in Fig. 5-5(b and c), H is maximized by making $w(x) = 0$ (in (b)) or 1 (in (c)). No intermediate value of $w(x)$ can be optimal, so the optimal control is bang-bang. When, however, a graph of $\mu(w)$ against $w(x)$ has the shape shown in Fig. 5-5(d), intermediate values of $w(x)$ can be optimal, as the broken line indicates.

The assumption that $\lambda(x)$ is positive can be justified as follows. Equations 5.5 show that $d\lambda(x)/dx$ equals $-\partial H/\partial P(x)$, which is necessarily negative if $\lambda(x)$ is small. Also, $\lambda(x)$ approaches zero as x approaches infinity. Hence $\lambda(x)$ must be positive at any finite age x.

Bang-bang control in this example means that once reproduction is started it happens at the maximum possible rate, even though this may result in very early death. The only situations (within the scope of the model) in which it can be better to reproduce more slowly and live longer, are ones in which mortality increases faster than in proportion to rate of reproduction ($d^2\mu/dw^2$ is positive, Fig. 5-5(d)).

How realistic is this model? It depends on the assumption that fast reproduction is apt to be followed by early death. Snell and King (1977) showed that this is true of a rotifer (*Asplanchna*). This animal reproduces parthenogenetically in suitable conditions, with no need for males. Newborn females were placed singly in small dishes of water and kept in carefully controlled conditions. Their offspring were removed and counted at 12 hour intervals, and the times of death of the mothers were recorded. It was found that individuals that produced fewer offspring per day tended to live longer than ones that reproduced faster. For instance, rotifers kept at 25°C were divided into two groups according to lifespan. The short-lived group lived on average 2.5 days and produced (while adult) an average of 4.2 offspring daily. The long-lived group lived 4.0 days and produced only 3.2 offspring daily. The points on a graph of $\mu(x)$ against $w(x)$ were rather scattered, but suggested a straight line (like Fig. 5-5a).

Many animals do not breed continuously but produce a series of discrete clutches of eggs, one after the other. If such an animal adopted the bang-bang strategy the first clutch would be as large as possible and the animal might be very likely to die before producing a second clutch. Bang-bang strategists might be likely to breed once only and then die, like Pacific salmon. These fish breed in rivers but spend much of their life in the sea. Breeding is preceded by an often long and strenuous journey up the river. This journey has to be made, however many or few offspring are produced at its end. It seems extremely likely that graphs of parental mortality against number of offspring, for such fish, would have shapes more or less like Fig. 5-5(b and c), so that the model would indicate a bang-bang strategy as optimal.

Botryllus schlosseri is a colonial ascidian found in rock pools on the shore of Cape Cod where there are two genetically distinct morphs (Grosberg, 1988). The life history starts in the same way in both cases: a larva settles and metamorphoses, then performs a series of cycles of budding to produce a colony composed of many interconnected zooids. Eventually, sexual reproduction occurs. Morph A grows very fast, each zooid budding off three new ones in each cycle. It produces one very big clutch of larvae (about 10 per zooid) and then dies. Morph B grows more slowly, producing only two buds per zooid in each cycle. It produces four or five clutches of larvae, but there are only about 3 larvae per zooid in each clutch. Both morphs are found throughout the year, but morph A is much more plentiful in the early part of the summer and morph B is more common in late summer.

The rate of increase r calculated from equation *5.16* is 40% greater for morph A than for morph B, so one might expect natural selection to eliminate the latter. However, A competes less well with another ascidian species which becomes plentiful in midsummer, and suffers severe mortality. Perhaps its faster growth and larger clutches leave it with less stamina for competition.

5.5 Sex ratios

Many species have approximately equal numbers of males and females, though they could produce just as many offspring in each generation if there were fewer males, each fertilizing several females. Why are there so many males?

Most animals are diploid (have a double set of chromosomes). They produce eggs and sperm which are haploid (have a single set of chromosomes) but which combine at fertilization so that the fertilized egg is diploid. Each son or daughter inherits half the father's genes and half the mother's genes. Each grandchild inherits one quarter of the genes of each grandparent.

Let a diploid population consist of n_m males and n_f females, which reproduce sexually to produce N offspring. The males produce on average N/n_m offspring and the females N/n_f. Let the cost of producing a son be C_m and the cost of producing a daughter C_f. (These costs may be assessed in terms of energy, or time, or some other currency.) An animal which invests C_m producing a son obtains on average N/n_m grandchildren so the number of grandchildren per unit investment is $N/n_m C_m$. Similarly, daughters yield $N/n_f C_f$ grandchildren per unit investment. Each grandchild inherits one quarter of the original animal's genes, whether these were transmitted through a son or through a daughter. Consequently, an animal with limited resources to invest may transmit more genes to subsequent generations by producing children of one sex rather than the other. Production of sons will be favoured by natural selection if

$$n_m C_m < n_f C_f$$

and production of daughters if

$$n_m C_m > n_f C_f$$

In either case the sex ratio will tend to change, in the course of evolution, until

$$n_m C_m = n_f C_f \tag{5.27}$$

Thus animals can be expected to devote equal proportions of their resources to producing sons and daughters (Fisher, 1930). For many species, there is no apparent difference in cost of production between sons and daughters, so the sexes should be produced in equal numbers.

This simple account ignores many complications which may arise, for instance if one sex matures faster than the other or if there is so little movement in the population that brothers compete mainly with each other for mates. It is nevertheless consistent with the observation that many species have sex ratios of about 1.

An extension of the same theory has been applied to the Hymenoptera, the ants, bees and wasps (Trivers and Hare, 1976). The extension was needed because one of the assumptions of the original theory is false for them. They are haplo-diploid: females develop from fertilized eggs and are diploid but males develop from unfertilized eggs and are haploid. A further complication is that many Hymenoptera are social, and their larvae are fed and cared for by sisters rather than parents.

Colonies of most species of social Hymenoptera resemble the wasp colonies discussed in section 5.2: they have just one queen and no reproducing male. The queen has stored in her body the sperm of (probably usually) just one male, with whom she copulated before founding the colony. He is the father of all her daughters, but her sons have no father because they develop from unfertilized eggs. Her offspring are cared for by (female) workers. Ant workers are typically sterile but some bee and wasp workers lay some eggs, which develop into males because they are not fertilized.

Many of the queen's offspring are workers or soldiers (sterile females that guard ant colonies). Others are reproductive males and females that leave the colony, when mature, to breed. In species whose workers lay no eggs, only the reproductives transmit her genes to future generations. Each reproductive, of either sex, carries half her genes. Natural selection will favour characteristics in the queen which tend to make the sex ratio satisfy equation *5.27*.

In this context, n_m and n_f are the numbers of reproductive males and females and C_m, C_f are their costs to the queen.

In practice, it seems most unlikely that the queen can control the sex ratio. The hundreds of workers which feed her offspring can probably frustrate any efforts she makes, for instance by neglecting larvae of one sex. Natural selection must favour genes which tend to make workers adjust the sex ratio, to a value which maximizes the number of the workers' genes which appear in subsequent generations. This may not be the same as the ratio indicated by equation 5.27.

If workers do not breed, they do not themselves transmit genes to subsequent generations. Nevertheless, the reproductives they care for are their brothers and sisters and share many of their genes, so the sex ratio of the reproductives affects the number of genes identical to the workers' genes, which appear in subsequent generations. Compare a female reproductive to a worker in the same colony. They have in common all the genes they inherited from their haploid father and (on average) half the genes they inherited from their diploid mother. Three quarters of their genes are in common. On the other hand a male reproductive has only one quarter of the workers' genes (half the genes she inherited from her mother). The male will transmit to each of his offspring only one third as many of the workers' genes, as the female transmits to each offspring. Equation 5.27 needs modification to take account of this. The appropriate equation is

$$3n_m C_m = n_f C_f \qquad (5.28)$$

Here n_m and n_f are the numbers of reproductive males and females in a population in which the sex ratio is controlled by the workers, and C_m and C_f are the costs *to the workers* of producing reproductive males and females. Since the workers feed the reproductives throughout larval life these costs will be assumed to be proportional to the dry masses of the adult reproductives.

Intact colonies of many species of ant have been dug up, and the reproductives have been counted and weighed. Male reproductives were up to eight times as common as females but individual

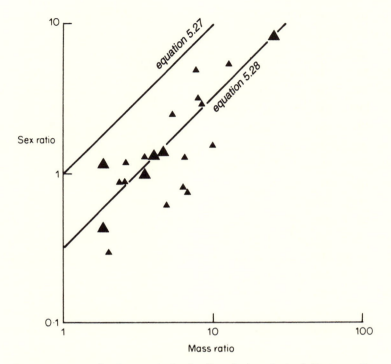

Figure 5-6. A graph of sex ratio (males/females) against adult mass ratio (females/males) for reproductives of 21 species of ant. All these species have just one queen in each colony, and none of them recruit slave workers of other species. The data were collected by Trivers and Hare (1976), who judged the points represented by large symbols to be the most reliable on the basis of criteria such as sample size.

females had up to 25 times the mass of males. Figure 5-6 shows that for any mass ratio C_f/C_m, the sex ratio n_m/n_f was generally fairly close to the value suggested by equation 5.28. Equation 5.27 fits the data much less well.

In some species of bee, a queen that dies is replaced by one of the workers which becomes sexually mature, mates, and lays eggs. The new queen is the workers' sister, not their mother, and there is no difference between their preferred sex ratio and hers. Mueller (1991) took advantage of this to test the theory. He removed the queen from some colonies of the bee *Augochlorella*, but not from others. He compared the sex ratios of reproductives hatching from eggs laid after the removals had been done. Colonies that retained the original queen had the expected female-biassed sex ratios, and those from which the queen had been removed had ratios somewhat biassed to males. If all the colonies had lost their original queens, both the workers and the new queens would be expected to prefer equal investment in male and female offspring, but if colonies that retain their queens are producing mainly females, those that have lost their queens may be expected to aim to balance this by producing mainly males.

The theory presented in this section predicts a value for the ratio $n_m C_m / n_f C_f$, but says nothing about the sex ratio n_m / n_f or about C_m and C_f. Macnair (1978) pointed out that the sizes of the two sexes (and so C_m and C_f) may have optimum values that arise in the same sort of way as the optimum size for eggs (section 5.1).

5.6 A battle of the sexes

Many vertebrate parents care for their eggs and young. For instance, birds incubate their eggs and feed their young. Sticklebacks (*Gasterosteus*) and many other bony fish guard their eggs. Caring for eggs and young increases their chances of survival but it uses time and energy which might otherwise be used to produce more fertilized eggs. A male might desert his mate as soon as he had fertilized her eggs, and go in search of another mate. In some circumstances this would increase his fitness if his original mate stayed to care for the eggs. Similarly, a female might be able to increase her fitness by deserting her mate if he remained. The balance of evolutionary advantage to each partner depends on the behaviour of the other, so the theory of games may be applicable.

Maynard Smith (1977) presented three games-theory models of parental behaviour. The one which seems most applicable to bony fish will be described. Let P_0, P_1 and P_2 be the probabilities of survival of unguarded eggs, eggs guarded by one parent and eggs guarded by two. It seems reasonable to assume that $P_0 \leqslant P_1 \leqslant P_2$. Let a male who deserts his mate immediately after fertilizing her

Table 5.1. The numbers of offspring produced by male and female animals which give or withhold parental care. Further explanation is given in the text.

	No. of offspring of male			No. of offspring of female	
	Female guards	Female deserts		Female guards	Female deserts
Male guards	nP_2	NP_1	Male guards	nP_2	NP_1
Male deserts	$nP_1(1 + p)$	$NP_0(1 + p)$	Male deserts	nP_1	NP_0

eggs have probability p of finding another mate. A female who deserts is much less likely to be able to breed again, in the same season, because of the time needed to grow a new batch of eggs. (In mature female fish, the eggs may be 20% or more of body mass.) It will be assumed for simplicity that she has no chance of breeding again. Also, a female who has just laid a very large batch of eggs may be weaker and less able to guard them than if she had laid a smaller batch. (Feeding may not be possible during guarding.) Let a female who is going to desert lay N eggs whereas one who stays to guard is restricted to n ($n \leqslant N$).

Both parents may guard the eggs, either one may guard or both may desert. The numbers of young produced by each parent in each case are shown in Table 5.1. There are four possible evolutionary stable strategies.

(i) Both sexes guard the eggs. This cannot be an ESS unless

$$P_2 \geqslant P_1(1 + p)$$

(or the male could increase his fitness by deserting) and

$$nP_2 \geqslant NP_1$$

(or the female could increase her fitness by deserting). This will generally be an ESS if two parents are much better than one ($P_2 \gg P_1$). This is more likely to be the case for insectivorous birds (which have to find food for their young) than for fish (which do not).

(*ii*) The male alone guards the eggs. This cannot be an ESS unless

$$P_1 \geqslant P_0(1 + p)$$

(or the male could increase his fitness by deserting) and

$$NP_1 \geqslant nP_2$$

(or the female could increase her fitness by guarding). This is likely to be the case if two parents are little better than one but one is much better than none ($P_2 \simeq P_1 \gg P_0$).

(*iii*) The female alone guards the eggs. This cannot be an ESS unless

$$P_1(1 + p) \geqslant P_2$$

(or the male could increase his fitness by guarding) and

$$nP_1 \geqslant NP_0$$

(or the female could increase her fitness by deserting). This may often be an alternative ESS to (*ii*) since both are likely to be ESSs if $P_2 \simeq P_1 \gg P_0$.

(*iv*) Both sexes desert. This cannot be an ESS unless

$$P_0(1 + p) \geqslant P_1$$

(or the male could increase his fitness by guarding) and

$$NP_0 \geqslant nP_1$$

(or the female could increase her fitness by guarding). This is likely to be an ESS if one parent is little better than none.

These arguments suggest that all four possibilities may be ESSs, in different circumstances. All four are realized among bony fishes, as the following examples show (Sterba, 1962).

(*a*) Angelfish (*Pterophyllum*, a fish often kept in tropical aquaria) lay their eggs on water plants. Both parents stay with the eggs, fanning water over them. They seem to assist hatching by chewing the eggs and they transfer the young larvae to shallow pits in the bottom.

(*b*) The male stickleback builds a barrel-shaped nest of plant fragments, and he alone guards the eggs that the female lays in it. He drives off any other fish or insects which may seem to threaten the eggs. The young hatch after one to two weeks but the male continues to guard them for some time thereafter. Throughout his period of guarding he does not feed.

(*c*) African freshwater fish of the genus *Haplochromis* have the habit known as mouthbrooding. The females take the newly-laid eggs into their mouths and keep them there until they hatch ten days or so later. After hatching the young are allowed out to feed, but they stay with their mother for several weeks, returning to her mouth at night and in times of danger. The mother cannot feed while the eggs are in her mouth but may start to feed again when the young are allowed out.

(*d*) Nearly all marine teleosts, except those that live on shores, lay eggs which float in the plankton and receive no parental care.

Blumer (1979) surveyed the 422 families of bony fishes and found that parental care occurs in 84 of them. Care by both sexes occurs in 21 families, by the male alone in 49 and by the female alone in 31. (In many of the families, different species have different habits.)

5.7 Parents and cuckolders

Bluegill sunfish (*Lepomis macrochirus*) live in lakes in eastern North America. Males adopt one or other of two reproductive strategies which we will call Parental and Cuckolder (Gross, 1991). Parental males do not mature until they are seven or eight years old. They then congregate in spawning areas where each prepares and guards a nest, a shallow depression in the lake bottom. Females lay their eggs in the nests, where the guarding males fertilize them by releasing sperm over them. The females leave, but the Parental males remain to care for the developing eggs.

Cuckolders mature much younger, when only two years old. They take advantage of their small size and agility to sneak into a nest where a female is laying and fertilize some of the eggs. They leave the cuckolded owner of the nest to care for their offspring as well as his.

Males are either Cuckolders or Parentals—Cuckolders do not mend their ways in later life. Only about 20% of two-year-old males become Cuckolders but, because so much mortality occurs between then and the age at which Parental males mature, over 80% of the mature male population are Cuckolders.

If Cuckolders were rare, the few that there were might do very well. If Parental males were rare, there would be very few nests for cuckolders to sneak into. Consequently, the optimum strategy for any particular male depends on what other males do, and game theory leads us to expect that evolution will find an ESS in which Cuckolders and Parental males have equal fitness. Similarly in the Hawk-Dove game (Table 4.3), the ESS gave Hawks and Doves equal scores.

Let the probability that a two-year-old male will still be alive at age x be $(P(x)$, and let the number of eggs he will fertilize at age x be $f(x)$. The total number of eggs he can be expected to fertilize in the course of his life (which we will use as our measure of fitness) is $\int_2^\infty P(x)f(x)\,dx$. At the ESS, the fitnesses of the two strategies must be equal. Using subscripts C for Cuckolder and P for Parental:

$$\int_2^\infty P_C(x)f_C(x)\,dx = \int_2^\infty P_P(x)f_P(x)\,dx = F \qquad (5.29)$$

Each of these integrals is the number of eggs that this year's two-year-old males will fertilize this year, plus the number that those that survive will fertilize next year, plus the number that survivors will fertilize the year after next, and so on. If the age distribution is stable, this is equal to the total number of eggs fertilized this year, by males of all ages. If the ESS has been established, equation 5.29 will apply. If we have both a stable age distribution and an ESS, and if N_C of this year's two-year-old males are Cuckolders, the number of eggs fertilized by Cuckolders of all ages this year will be $N_C F$. Similarly, if N_P of this year's two-year-old males become Parentals, the number of eggs fertilized by Parental males this year will be $N_P F$. Thus, of this year's eggs, a fraction $N_C F/(N_C F + N_P F) = N_C/(N_C + N_P)$ will be fertilized by Cuckolders. In other words, if the ESS has been established, the fraction of eggs fertilized by cuckolders will equal the fraction of two-year-old males who become Cuckolders.

Female sunfish lay their eggs in a large number of very small batches, about twelve eggs in each batch. SCUBA divers observing nests carefully were able to see how many of these batches were fertilized by the Parental male and how many by Cuckolders. They observed nests attended by different numbers of Cuckolders and confirmed as expected that a Cuckolder's mating success depends on how many other Cuckolders there are around the same nest. Also, they estimated from their observations that the proportion of eggs fertilized by Cuckolders in the population as a whole was approximately equal to the proportion of two-year-old males who become Cuckolders. This result is consistent with the hypothesis that an ESS had been established.

6

Dangers and difficulties

OPTIMIZATION THEORY has become a very popular tool in biology, and it seems to be a powerful one, but even the best tools do good work only when properly used. This chapter is about some of the dangers, difficulties and limitations of optimization theory.

6.1 What is optimized?

The first problem in using optimization theory is to decide what should be optimized. Natural selection tends to maximize fitness. (This is an axiom: it is implicit in the definitions of the words.) Therefore, it might be argued, the criterion in all our optimization models should be the maximization of fitness.

This would lead to great difficulties, in many cases. Consider as an example the discussion of the compound eyes of insects in section 2.3. The problem was posed as the minimization of visual resolution (i.e. minimization of the angle between two objects that can just be seen as separate). To re-formulate it as a problem of maximizing fitness, it would be necessary to establish a quantitative relationship between resolution and fitness. I would find this extraordinarily difficult to do, but it seems obvious enough that as a general rule, improved resolution will increase fitness by making animals better able to find food, avoid predators etc. Rather than complicate the mathematics by introducing a hypothetical and possibly spurious equation relating resolution to fitness, it seemed better to investigate how resolution could be optimized.

In most of the discussions of movement in Chapter 3, the optimization criterion is minimization of power requirements. In the discussions of feeding in Chapter 4, the criterion is maximization of the rate of intake of food. It might have been possible to translate these discussions, so that fitness was being maximized directly. A quantitative relationship has been postulated for fish, showing how adaptations that reduce energy expenditure or increase food intake should increase fitness, if various assumptions are valid (Alexander, 1967). Similar equations based on similar assumptions could be written for other animals and incorporated in the models in Chapters 3 and 4, but I do not consider that this would be sensible. It would make the models more complicated and it would add little to their value because the extra link in the chain of argument would be a weak one, in the present state of knowledge.

Fitness is more directly related to reproductive strategy, than to visual resolution, energy expenditure or food intake. It therefore seemed sensible in Chapter 5, to use fitness itself as the criterion for optimization.

It may be helpful to use mathematical language to clarify some of the assumptions behind models that seek optima for quantities other than fitness. It is assumed that fitness Φ can be regarded as a function of a large number of other quantities that will be called fitness components and given symbols y_1, y_2 etc.

$$\Phi = F(y_1, y_2, y_3 \ldots)$$

For instance, y_1 might be visual resolution, y_2 might be the power required for walking, y_3 might be rate of food intake and so on. It is also assumed that for every y, $\delta\Phi/\delta y$ is either always positive or always negative. For instance it is assumed that reduced power requirement and increased food intake always increase fitness, so $\delta\Phi/\delta y_2$ is always negative and $\delta\Phi/\delta y_3$ is always positive.

The fitness components are functions of another set of variables that will be called characters and given the symbols z_1, z_2 etc. Thus visual resolution (y_1) is a function of ommatidial diameter (z_1), and power requirement for walking (y_2) is a function of stride frequency (z_2), duty factor (z_3) and shape factor (z_4).

$$y_1 = f(z_1)$$
$$y_2 = f(z_2, z_3, z_4)$$
$$y_3 = f(z_5, z_6)$$

and so on, where each f represents a different function.

The ultimate aim is to discover the values of z_1, z_2 etc. that maximize Φ. This has been attempted indirectly by seeking the value of z_1 that optimizes y_1, the values of z_2, z_3, and z_1 that optimize y_2, and so on. (Optimize here means maximize or minimize, whichever is required to maximize fitness.) This procedure is apt to fail if several fitness components are functions of the same character. For instance if

$$y_4 = f(z_7, z_8)$$
$$\text{and } y_5 = f(z_8, z_9)$$

different values of z_8 may be required to optimize y_4 and y_5.

This sort of difficulty arose in the discussion of eggshells (section 2.4). Increasing porosity of the eggshell tends to reduce fitness by increasing water loss from the egg, but it also tends to increase fitness by allowing the partial pressure of oxygen in the egg to increase. If water loss and oxygen partial pressure had been regarded as separate fitness components, two different 'optimum' porosities would have been calculated. Similarly in the discussion of hungry, thirsty doves (section 4.6), hunger and thirst could not be treated as separate fitness components because drinking prevents simultaneous eating, and vice versa. The approach adopted in each of these cases was to define a composite fitness component. Equation 2.9 defines a fitness component that takes account of both water loss and oxygen partial pressure, and equation 4.9 defines one that takes account of both hunger and thirst. In both cases the procedure is rather unsatisfactory because it is largely arbitrary: different, equally plausible equations could have been written (see Sibly and McFarland's, 1976, discussion of the hunger and thirst problem).

Fortunately, the conclusions from optimization models are often altered only a little by radical changes in the mathematical form of fitness components. For instance, the hunger and thirst model involved a fitness component

$$x_1^a + b x_2^a,$$

where x_1 is food deficit, x_2 is water deficit and a and b are constants. It was shown that this is reduced fastest by the strategy of eating or drinking, whichever is necessary to bring the (x_1, x_2) to the frontier indicated by the line in Fig. 4-6(a), and then moving down the frontier by eating and drinking alternately. This is the case, whatever the values of a and b, provided only that $a > 1$ and $b > 0$: the only effect of changing a and b is to move the frontier. The optimum strategy remains the same even if the fitness component y, has an entirely different form such as

$$y = \exp(x_1) + \exp(cx_2)$$

The only requirement is that $\partial^2 y / \partial x_1^2$ and $\partial^2 y / \partial x_2^2$ should always both be positive. The same sort of generality was made explicit in the discussion of egg size (section 5.1): no equations were written for the graphs in Fig. 5-1, and the conclusions depended only on whether they were convex or concave.

Composite fitness components are sometimes avoided by neglecting a possible fitness component as unimportant. For instance, in the discussion of mammalian gaits (sections 3.3 and 3.4), economy of energy was taken as the fitness component and steadiness was ignored. In the discussion of tortoise gaits (section 3.5), however, steadiness was taken as the fitness component and economy was ignored. These choices were not arbitrary (it was argued that steadiness becomes increasingly important as speed decreases) but nevertheless involved subjective judgment. Some readers might have preferred to define a composite function taking account of both steadiness and economy, and use it as the fitness component in both cases. They would have found it difficult to decide what function should be used.

A similar problem arose in the discussion of moose diets (section 4.2). Is the appropriate fitness component daily energy intake (which should be maximized), time spent feeding (which should be minimized) or a composite of the two? Belovsky (1978) did not attempt to answer the question from first principles, but calculated optimum diets for energy maximization and time minimization. He found that the actual diet was more like the former.

6.2 What is possible?

The next problem, after deciding what is likely to be optimized, is to decide what structures or strategies are possible, and what constraints apply. If no such limitations were recognized it would have to be concluded that optimum structure would make bones unbreakable and without mass, and an optimum life-history would involve immortality and infinite fecundity.

Problems usually have to be posed in ways that allow only a limited range of answers. For instance, the discussion of bones (section 2.1) does not ask the general question, what is the best shape for the cross-section of a bone? Instead, it seeks the best proportions for a tubular bone of circular cross-section. The tibias of mammals are far from circular in section and the humerus of some birds are reinforced by internal buttresses. It would have been much more difficult to formulate an optimization model that permitted structures like these. The models of animal conflicts in section 4.10 are even more restrictive: only five specific strategies are considered. Several other strategies were also considered by Maynard Smith and Price (1973), but it is entirely possible that there is a strategy that is an ESS against all of them, that no one has thought of.

Constraints are often not explicit in optimization models, but implicit in assumptions. For instance, it is assumed in the discussion of eyes (section 2.3) that the range of wavelengths to which the eye is sensitive cannot be made shorter, to improve resolution.

It can be difficult to decide which of several alternative constraints take effect. For instance, there must be a limit to the rate at which a queen wasp can lay eggs and there must also be a limit to the rate at which her workers can collect food for the larvae. Which constraint is effective in limiting the rate of growth of a wasp colony? It is assumed in section 5.2 that the rate of collection is limiting (consistent with the increasing rate of egg production through the season), but the constraint has to be defined more precisely. Is the rate limited by the number of workers (in which case it would presumably be proportional to the number) or by the amount of potential food in the environment (in which case it might be independent of the number of workers) or by a combination of both? The model in section 5.2 assumes the first of these alternatives but Macevitz and Oster's (1976) original version of the model is more elaborate, allowing for effects of food availability. The two models lead to very similar conclusions for hornets.

6.3 What can go wrong?

The classical procedure in science is to formulate a hypothesis and then carry out an experiment that is capable of disproving it. If the results agree with the hypothesis, it is retained but if not, it is rejected. Optimization theory is used to generate hypotheses about zoology which are then tested by observation or experiment. For instance, the discussion of safety factors (section 2.2) produced the hypothesis that bones have tended to evolve so as to minimize the cost specified by equation *2.4*. This hypothesis has not yet been tested. The marginal value theorem (section 4.3) generated hypotheses that particular species of animal, depleting patches of food by feeding on them, would tend to move on to the next path at the optimum time indicated by the theorem. Such hypotheses have survived several tests including the one with ladybird larvae illustrated in Fig. 4-3.

Sometimes tests like these give results at variance with the hypothesis. Several kinds of explanations are possible. One possibility is that an error has been made in the mathematics that generated the hypothesis. Another is that the animals have failed to achieve the optimal structure or behaviour either because natural selection for the structure or behaviour is still in progress and equilibrium has not been reached, or because appropriate mutations have not appeared. The explanation that evolution has not got there yet should be used cautiously. The explanation that appropriate mutations are not available may often apply when the optimal solution is a difficult one, but the complexity of what has evolved is often amazing. I find it astonishing that tits are able to come so near to the optimal solution to the two-armed bandit problem (section 4.5), since that solution can only be reached by complex mathematics. The birds are presumably using some relatively simple algorithm: I do not for a moment believe that they can work out Bayesian probabilities.

Discrepancies between hypothesis and observation are often due to unrealistic assumptions implied by the hypothesis. For instance, it had been shown that the gait that minimizes unsteadiness for a walking quadruped is the one illustrated in Fig. 3-8(a). This gait involves no unwanted displacements whatever. It might have been expected that tortoises would walk like this, but tortoises actually use the gait shown in Fig. 3-8(c). This suggested that gait (a) might be impossible for tortoises. Tortoises have slow muscles which are very economical of energy, but there was no constraint in the original model to limit rates of change of force. A new mathematical model was devised, incorporating the constraint that forces must rise and fall slowly as shown in Fig. 3-4(a). This led to the conclusion that the gait shown in Fig. 3-8(b) is optimal. This is more like the actual gait than gait(a) is, but not identical with it. It was suspected that the constraint had been unrealistically severe, so it was relaxed a little to allow skewed force patterns. The modified model indicated as optimal a gait that is extremely like the observed one.

You should be cautious about tinkering with models to make them fit existing data. It is always dangerous unless the modification is theoretically plausible and the modified hypothesis is tested by further observation or experiment. I think the dangers were avoided in this particular case. The modification was plausible. It is consistent with our limited knowledge of tortoise muscle physiology. It was the obvious modification to try since it allows faster muscle action by (in effect) adding a first harmonic to a fundamental. It was tested by calculating optimum relative phases and degree of skewness and comparing them with observation. The actual relative phases were already known but the degree of skewness was calculated from force records at this stage.

Another discrepancy between observation and original hypothesis arose in a study of egrets (*Casmerodius*) in Florida. These birds

fish like other herons, but quite frequently attack another bird
that is fishing nearby and attempt to steal a fish from it (Kushlan,
1978). This suggested two hypotheses: first, that egrets attempt
robbery to minimize the time they have to spend feeding, and sec-
ondly (more specifically) that rate of food energy intake while robbing is greater than while fishing. The second hypothesis seemed
to be a necessary corollary of the first. Robbing attempts often
fail, however, and it was estimated from field observations that
egrets obtained food energy at mean rates of 0.47 W while fishing
and only 0.15 W while attempting robbery. This seemed to dis-
prove the second hypothesis and (by implication) the first one.
Later it was realized that attackers often displace victims from
their feeding places (Caldwell, 1980). This suggested a new second
hypothesis, that attacking birds increase their longer-term rate of
food intake by capturing good fishing places. This was tested in
Panama, by comparing the numbers of fish caught in five-minute
periods immediately before and after an attack. Attackers caught
fish significantly faster, after a successful attack.

6.4 Criticisms

This book shows that the optimization approach is a powerful
tool for biologists. Nevertheless, the approach has been fiercely
criticized, most famously by Gould & Lewontin (1979) in their ec-
centrically titled paper, "The spandrels of San Marco and the Panglossian paradigm." One of the points they make is that biologists
have often looked for optimization, without taking into account
the constraints that an animal's ancestry impose on its evolution.
A mollusc cannot be expected to evolve the same structure as a
fish. The next few paragraphs amplify this point.

A function may have several maxima and minima, like peaks and
valleys in a mountain range. In such cases the highest maximum
is called the global maximum and the others are local maxima.
Evolution may fail to reach a global optimum because a species
evolves to a local optimum. Any small move away from the local
optimum would be disadvantageous, so the evolving species may

get trapped there. An analogy may make this clearer. Suppose you start walking in a mountain range, going always uphill. You will eventually reach a peak, but not necessarily the highest peak.

Something like this seems to have happened in the evolution of squids. With their streamlined bodies and well-developed muscles they seem beautifully adapted for swimming, but they compare very badly with fishes in swimming ability. Measurements of swimming speed and oxygen consumption of squid and trout of equal mass showed that the squid swam less fast than the trout and nevertheless used more energy (Webber & O'Dor, 1986). The explanation seems to be that they swim by jet propulsion, which is less efficient than the trout's technique of swimming by beating its tail. Squid may have reached a local optimum in the evolution of their swimming mechanism, but they would swim better if they could evolve fish-like tails. It seems most unlikely that they ever will, because the early stages of any change from jet propulsion to tail-beating would surely be less good than the system they have.

Another criticism voiced by Gould & Lewontin (1979) is that practitioners of the optimization approach are so blind to alternative kinds of explanation that if one optimization story fails they will simply substitute another—any plausible story will do. This criticism seems to grow from a misunderstanding of the role of optimization arguments (see Parker & Maynard Smith, 1990). Their proper function is not to prove that organisms are optimal, but to improve our understanding of specific cases of optimization by asking what particular sets of assumptions should lead us to expect. If the predictions of an optimization theory are very different from what we observe in nature, that suggests either that we have misunderstood an adaptation or that we have overlooked a constraint that arises, perhaps, from the animal's ancestry. We should try another theory, starting from different assumptions.

7

Mathematical summary

THIS FINAL chapter is a brief summary of the methods of finding optima that have been used in this book. There are many books on optimization theory that describe the same methods more fully. I like the one by Koo (1977), but some readers may prefer other books.

An optimum may be either a maximum or a minimum, but the problem of finding a minimum is easily converted to one of finding a maximum; minimizing a function Φ is the same as maximizing $-\Phi$. Problems are therefore presented in this chapter only as problems of maximization. Different kinds of problem are considered in turn.

7.1 Maxima of functions of one variable

The standard problem of this kind is:

choose x so as to maximize Φ

where $a \leqslant x \leqslant b$ and $\Phi = f(x)$

The aim is to find the value of x, within the specified range, that gives Φ, a function of x, its largest value. Several possible situations are shown in Fig. 7-1.

(a) In Fig. 7-1(a), Φ has its maximum value when x has the value $\hat{x}$, which is greater than a but less than b. This is described as an interior maximum. The value of $\hat{x}$ can be found by using the conditions

when $x = \hat{x}, d\Phi/dx = 0$ and $d^2\Phi/dx^2$ is negative.

These conditions are explained in section 1.2, but there is a complication which is not explained there. In some cases $d^2\Phi/dx^2$ is

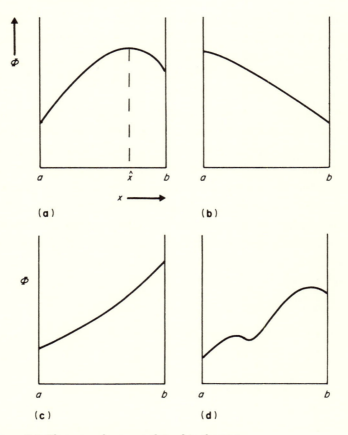

Figure 7-1. These graphs are explained in the text.

zero at the maximum. To cover such cases the conditions can be written

when $x = \hat{x}$, $d^n\Phi/dx^n$ (for $n = 1, 2, 3$ etc.) is zero for all values of n up to some even-numbered value, for which it is negative.

For example, $\Phi = -x^4$ has a maximum at $x = 0$. In this case $d\Phi/dx = -4x^3$, $d^2\Phi/dx^2 = -12x^2$ and $d^3\Phi/dx^3 = -24x$, which are all zero where $x = 0$, but $d^4\Phi/dx^4 = -24$, which is negative.

Another complication is that there may be several maxima in the range (Fig. 7-1d).

(b) If $d\Phi/dx$ is negative throughout the range $a \leqslant x \leqslant b$ there is no interior maximum (Fig. 7-1b). The maximum occurs at the lower edge of the range where $x = a$.

(c) If $d\Phi/dx$, is positive throughout the range the maximum occurs when $x = b$ (Fig. 7-1c).

The problems of finding optimum speeds for aircraft (section 1.3) and optimum conductance for eggshells (section 2.4) led to interior optima for functions of one variable. The problem of optimum worm selection (section 4.1) also involved a function of one variable, but the optimum occurred at the extreme of the range of possible values. The problem of bounding flight (section 3.1) led to an interior optimum only at higher speeds: in slow flight, equation 3.5 implies (wrongly) that the bird should flap its wings all the time.

7.2 Maxima of functions of several variables

Choose values within specified ranges for the variables x_1, x_2, x_3 etc. so as to maximize $\Phi = f(x_1, x_2, x_3,$ etc.$)$.

If there is an interior maximum at $(\hat{x}_1, \hat{x}_2, \hat{x}_3 \ldots)$, then when $x_1 = \hat{x}_1, x_2 = \hat{x}_2$ etc., the partial derivatives $\partial\Phi/\partial x_1$, $\partial\Phi/\partial x_2$, $\partial\Phi/\partial x_3$, etc. are all zero. A point that satisfies this condition may be a maximum or a minimum or neither. The formal criterion for deciding which is complicated (see Koo, 1977).

The problems of finding the line of best fit (section 1.4) and of choosing the best speed and leg angle for a jump (section 3.2) involved optimizing functions of two variables. In the case of jumping the functions were too complicated to be differentiated and the maxima were found by plotting graphs. Figure 3-2(d) shows an interior maximum and Fig. 3-2(c) shows a case where the maximum is at the edge of the range.

7.3 Maxima of functions with constraints

Choose values within specified ranges for the variables x_1 and x_2 so as to maximize $\Phi = f(x_1, x_2)$, subject to the constraint $g(x_1, x_2) = 0$ (f and g both represent functions).

Any interior maximum can be located by defining the function

$$L = f(x_1, x_2) + \lambda.g(x_1, x_2)$$

and differentiating it with respect to x_1 and x_2. The symbol λ represents an unknown constant, the Lagrange multiplier. If there is an interior maximum at $\hat{x}_1, \hat{x}_2$, then when $x_1 = \hat{x}_1$ and $x_2 = \hat{x}_2$

$$\partial L/\partial x_1 = 0, \partial L/\partial x_2 = 0 \quad \text{and} \quad g(x_1, x_2) = 0$$

These three equations must be solved simultaneously to determine x_1, x_2 and λ. A point that satisfies this condition may be a maximum or a minimum or neither: the criteria that say which are given in books such as Koo (1977). These books also explain how the method can be applied to problems with more than two unknowns, and several constraints.

This method was used to solve problems about food cans in section 1.5, but has not been applied to biological problems in this book. It was introduced because it is related to the Pontryagin method (section 5.2).

7.4 Linear programming

This is a method that can be applied to a special class of optimization problem with constraints.

Choose x_1, x_2, x_3 etc. so as to maximize the function $\Phi = a_1x_1 + a_2x_2 + a_3x_3 + \cdots$ subject to the constraints

$$b_1x_1 + b_2x_2 + b_3x_3 + \cdots \leqslant b_0$$
$$c_1x_1 + c_2x_2 + c_3x_3 + \cdots \leqslant c_0 \text{ etc.}$$

and also $x_1 \geqslant 0, x_2 \geqslant 0, x_3 \geqslant 0$, etc.

In this problem $a_1, a_2 \ldots b_1, b_2$, etc. represent constants. Notice that Φ and the constraints are all linear functions of x_1, x_2, x_3 etc; that is to say, these variables appear only as first powers (not squares, cubes, etc.) multiplied by constants.

If there are only two variables (x_1 and x_2), a graph can be drawn with axes representing x_1 and x_2, and the constraints can be drawn on the graph as straight lines (Fig. 4-2). The lines form a polygon enclosing all points (x_1, x_2) that satisfy the constraints. The maximum always occurs at a corner of the polygon.

If there are three variables, the constraints can be represented as planes in three-dimensional space. These form a polyhedron (a plane-faced solid) enclosing all (x_1, x_2, x_3) that satisfy the constraints. The maximum occurs at a corner of the polyhedron. If there are n variables ($n > 3$) the constraints are represented by hyperplanes in n-dimensional space, which is rather hard to imagine. The maximum occurs at a corner of the hyperpolyhedron.

If there are just two variables the maximum can be found by drawing a graph like Fig. 4-2. If there are three or more it can be found by evaluating Φ for all the corners of the polyhedron or hyperpolyhedron and seeing which value is the largest. If there are a lot of constraints, this is a long job, but there is a systematic method for finding the maximum without examining obviously hopeless corners. This method (the Simplex method) is explained in books such as Koo (1977).

The problem of the moose's diet (section 4.2) is a problem in linear programming.

7.5 Maximum of the smaller of two alternatives

$\Phi = f(x)$ or $g(x)$, whichever is the smaller for the particular value of x. Choose x in the range $a \leqslant x \leqslant b$ so as to maximize Φ.

Several different situations have to be considered.

(a) If $f(x) < g(x)$ for all values of x within the range, the problem is simply to maximize $f(x)$ (Fig. 7-2a).

(b) If df/dx is positive and dg/dx negative, for all values of x in the range, if there is a value of x for which $f(x) = g(x)$, then this value $\hat{x}$ is the one that maximizes Φ (Fig. 7-2b).

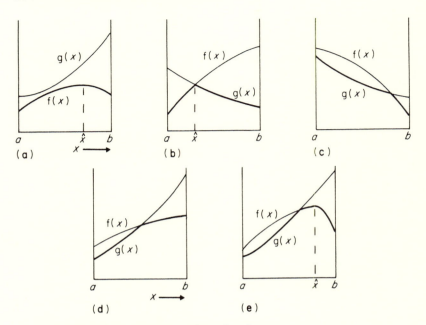

Figure 7-2. These graphs are explained in the text.

(*c*) If d*f*/d*x* and d*g*/d*x* are both negative throughout the range, the maximum occurs when *x* = *a* (Fig. 7-2c).

(*d*) If, however, they are both positive throughout the range, the maximum occurs when *x* = *b* (Fig. 7-2d).

(*e*) There are also other possibilities such as the one shown in Fig. 7-2(e).

The minimization problem corresponding to the problem just described, is to minimize the larger of two alternatives. Problems of this kind arose in the discussions of compound eyes (section 2.3) and air-filled bones (section 2.1). The problem of territory size (section 4.6), however, was one of maximizing the smaller of two alternatives.

Problems involving more than two alternatives can be solved in essentially the same ways as those that involve just two.

7.6 Calculus of variations

Choose values of the state variable $x(t)$ for all values of t from t_1 to t_2, so as to

$$\text{maximize } \Phi = \int_{t_1}^{t_2} f[x(t), \dot{x}(t), t].dt$$

Here $\dot{x}$ means dx/dt and f is a function of $x(t)$, $\dot{x}(t)$ and t. The required function $x(t)$ must satisfy the Euler equation

$$\frac{\partial f}{\partial x} - \frac{d}{dt}\left(\frac{\partial f}{\partial \dot{x}}\right) = 0$$

for all values of t from t_1 to t_2. The method can also be applied to problems involving several state variables. In such cases there are as many Euler equations as state variables.

The problems of the shortest line (section 1.6) and the hungry bird (section 4.6, second part) were solved by using the calculus of variations.

7.7 The Pontryagin method

Choose values of the control variable $u(t)$ for all values of t from t_1 to t_2 so as to

$$\text{maximize } \Phi = \int_{t_1}^{t_2} f[x(t), u(t), t]\, dt$$
$$\text{where } d[x(t)]/dt = g[x(t), u(t), t]$$

One of the conditions that must be satisfied by the solution is that for every value of t from t_1 to t_2, $u(t)$ must have the value that maximizes the Hamiltonian

$$H = f[x(t), u(t), t] + \lambda(t)g[x(t), u(t), t],$$

in which $\lambda(t)$ is a function of t such that

$$d\lambda/dt = -\partial H/\partial x$$
$$\text{and } \lambda(t_2) = 0$$

The same method can be used to tackle problems involving several state variables, but an equal number of functions λ_1, λ_2 etc. are then required. A slight modification allows the method to be used also for problems in which Φ has a non-integral term as well as the integral (Koo, 1977).

Three problems of finding optimum life histories were tackled by the Pontryagin method (sections 5.2, 5.3 and 5.4).

7.8 Dynamic programming

Dynamic programming is another method of finding optimum sequences of choices. It involves working backwards by stages from the end point. It is explained in section 4.7, where it is applied to a bird that needs to find enough food before nightfall.

7.9 Evolutionarily stable strategies

Consider a contest between animals in which several strategies are possible, and are adopted by different animals. Assume that the strategies adopted by particular animals are determined genetically. A strategy is described as evolutionarily stable if, in a population in which nearly all the individuals adopted it, none of the rare alternative mutant strategies would be favoured by natural selection.

In a particular case, let the only possible strategies be I and J, and let the payoffs be as shown in Table 7.1. The strategy I is an evolutionarily stable strategy (ESS) if

$$\text{either } \alpha > \beta$$
$$\text{or } \alpha = \beta \text{ and } \gamma > \delta.$$

Also, J is an ESS if

$$\text{either} \quad \delta > \gamma$$
$$\text{or} \quad \delta = \gamma \text{ and } \beta > \alpha$$

It is possible for both strategies to be ESSs, in which case the one

Table 7.1.

In a contest against		I	J
this animal obtains	$\{I$	α	γ
the payoff shown	$\{J$	β	δ

that becomes established under natural selection depends on the initial composition of the population.

Further explanation of ESSs is given by Maynard Smith (1982). The concept of the ESS is used in this book, in the sections about animal contests (4.10) and parental care (5.6).

7.10 Catastrophe theory

Some of the rudiments of catastrophe theory were explained in sections 2.2 (about factors of safety for bones) and 3.4 (about gait changes). Useful information about the numbers of possible kinds of catastrophe is given in the final paragraph of section 3.4. Further information is given by Saunders (1980).

References _____

Alexander, R. McN. 1967. *Functional design in fishes.* London: Hutchinson.
———. 1974. The mechanics of jumping by a dog (*Canis familiaris*). *J. Zool.* (London) **173**:549-73.
———. 1975. Evolution of integrated design. *Am. Zool.* **15**:419-25.
———. 1980. Optimum walking techniques for quadrupeds and bipeds. *J. Zool.* (London) **192**:97-117.
———. 1981. Factors of safety in the structure of animals. *Sci. Prog.* **67**:109-30.
———. 1983. *Animal mechanics.* 2d ed. Oxford: Blackwell.
———. 1989. Optimization and gaits in the locomotion of vertebrates. *Physiol. Rev.* **69**:1199-1227.
———. 1990. Optimum take-off techniques for high and long jumps. *Phil. Trans. Roy Soc. B* **329**:3-10.
———. 1991. Optimization of gut structure and diet for higher vertebrate herbivores. *Phil. Trans. Roy. Soc. B* **333**:249-55.
———. 1992. A model of bipedal locomotion on compliant legs. *Phil. Trans. Roy. Soc. B* **338**:189-98.
———. 1993. The relative merits of foregut and hindgut fermentation. *J. Zool.* (London) **231**:391-401.
Alexander, R. McN., & A. S. Jayes. 1978. Vertical movements in walking and running. *J. Zool.* (London) **185**:27-40.
———. 1980. Fourier analysis of forces exerted in walking and running. *J. Biomechan.* **13**:383-90.
Ar, A., C. V. Paganelli, R. B. Reeves, D. G. Greene, & H. Rahn. 1974. The avian egg: Water vapour conductance, shell thickness and functional pore area. *Condor* **76**:153-58.
Barlow, H. B. 1952. The size of ommatidia in apposition eyes. *J. Exp. Biol.* **29**:667-74.
Bell, G. 1980. The costs of reproduction and their consequences. *Am. Nat.* **116**:45-76.
Belovsky, G. E. 1978. Diet optimization in a generalist herbivore: The moose. *Theoret. Popn. Biol.* **14**:105-34.
Blumer, L. S. 1979. Male parental care in the bony fishes. *Q. Rev. Biol.* **54**:149-61.
Brown, M. E. 1957. Experimental studies on growth. In *The Physiology of Fishes.* Volume 1. London: Academic Press.
Caldwell, G. S. 1980. Underlying benefits of foraging aggression in egrets. *Ecology* **61**:996-97.
Caraco, T., S. Martindale, & T. S. Whittam. 1980. An empirical demonstration of risk-sensitive foraging preferences. *Anim. Behav.* **28**:820-30.

Carpenter, F. L., D. C. Paton, & M. A. Hixon. 1983. Weight gain and adjustment of feeding territory size in migrant hummingbirds. *Proc. Nat. Acad. Sci.* **80**:7259-63.

Charnov, E. L. 1976a. Optimal foraging: Attack strategy of a mantid. *Am. Nat.* **110**:141-51.

———. 1976b. Optimal foraging: The marginal value theorem. *Theoret. Popn. Biol.* **9**:129-36.

———. 1993. *Life history invariants. Some explorations of symmetry in evolutionary ecology.* Oxford: Oxford University Press.

Clutton-Brock, T. H., S. D. Albon, R. M. Gibson, & F. E. Guinness. 1979. The logical stag: Adaptive aspects of fighting in red deer (*Cervus elaphus L.) Anim. Behav.* **27**:211-25.

Colgan, P. W., W. A. Mowell, & M. W. Stokes. 1981. Spatial aspects of nest defence by pumpkinseed sunfish *(Lepomis gibbosus)*: Stimulus features and an application of catastrophe theory. *Anim. Behav.* **29**:433-42.

Cook, R. M., & B. J. Cockrell. 1978. Predator ingestion rate and its bearing on feeding time and the theory of optimal diets. *J. Anim. Ecol.* **47**:529-47.

Cowie, R. J. 1977. Optimal foraging in great tits *(Parus major). Nature* (London) **268**:137-39.

Currey, J. D., & R. McN. Alexander. 1985. The thickness of the walls of tubular bones. *J. Zool.* (London) **206**:453-68.

Davies, N. B. 1978. Territorial defence in the speckled wood butterfly *(Pararge aegeria)*: The resident always wins. *Anim. Behav.* **26**:138-47.

Davis, T. A., M. F. Platter-Reiger, & R. A. Ackerman. 1984. Incubation water loss by pied-billed grebe eggs: Adaptation to a hot, wet nest. *Physiol. Zool.* **57**:384-91.

Fisher, R. A. 1930. *The genetical theory of natural selection.* Oxford: Oxford University Press.

Forsyth, A. B., & R. J. Robertson. 1975. K reproductive strategy and larval behaviour of the pitcher plant sarcophagid fly, *Blaesoxipha fletcheri. Can. J. Zool.* **53**:174-79.

Geist, V. 1966. The evolution of horn-like organs. *Behaviour* **27**:175-214.

Goss-Custard, J. D. 1977. Optimal foraging and the size selection of worms by redshank, *Tringa totanus*, in the field. *Anim. Behav.* **25**:10-29.

Gould, S. J., & R. C. Lewontin. 1979. The spandrels of San Marco and the Panglossian paradigm: A critique of the adaptationist programme. *Proc. Roy. Soc. B* **205**:581-98.

Green, R. F. 1980. Bayesian birds: A simple example of Oaten's stochastic model of optimal foraging. *Theoret. Popn. Biol.* **18**:244-56.

Grosberg, R. K. 1988. Life history variation within a population of the colonial ascidian *Botryllus schlosseri. Evolution* **42**:900-920.

Gross, M. R. 1991. Evolution of alternative reproductive strategies: Frequency dependent sexual selection in male bluegill sunfish. *Phil. Trans. Roy. Soc. B* **332**:59-66.

Harper, D.G.C. 1982. Competitive foraging in mallards: 'Ideal free' ducks. *Anim. Behav.* **30**:575-84.

Hixon, M. A., F. L. Carpenter, & D. C. Paton. 1983. Territory area, flower density, and time budgeting in hummingbirds: An experimental and theoretical analysis. *Am. Nat.* **122**:366-91.

Houston, A. I., & J. M. McNamara. 1988. A framework for the functional analysis of behaviour. *Behav. Brain Sci.* **11**:117-54.

Hughes, R. N., ed. 1993. *Diet selection: An interdisciplinary approach to foraging behaviour.* Oxford: Blackwell.

Ishay, J., H. Bytinski-Salz, & A. Shulov. 1967. Contributions to the bionomics of the oriental hornet (*Vespa orientalis* Fab.) *Israel J. Entomol.* **2**:45-106.

Jayes, A. S., & R. McN. Alexander. 1980. The gaits of chelonians: Walking techniques for very low speeds. *J. Zool.* (London) **191**:353-78.

Koo, D. 1977. *Elements of optimization: With applications in economics and business.* New York: Springer.

Krebs, C. J., ed. 1985. *Ecology. The experimental analysis of distribution and abundance.* 3d ed. New York: Harper & Row.

Krebs, J. R., & N. B. Davies. eds. 1993. *An introduction to behavioural ecology.* 3d ed. Oxford: Blackwell.

Krebs, J. R., J. T. Erichsen, M. E. Webber, & E. L. Charnov. 1977. Optimal prey selection in the great tit (*Parus major*). *Anim. Behav.* **25**: 30-38.

Krebs, J. R., A. Kacelnik, & P. Taylor. 1978. Test of optimal sampling by foraging great tits. *Nature* (London) **275**:27-31.

Kushlan, J. A. 1978. Nonrigorous foraging by robbing agrets. *Ecology* **59**:649-53.

Lack, D. 1946. *The life of the robin.* London: Witherby.

Larkin, S., & D. McFarland. 1978. The cost of changing from one activity to another. *Anim. Behav.* **26**:1237-46.

León, J. A. 1976. Life histories as adaptive strategies. *J. Theor. Biol.* **60**:301-35.

Lighthill, M. J. 1977. Introduction to the scaling of aerial locomotion. Pages 365-404 in *Scale effects in animal locomotion*, ed. T. J. Pedley. London: Academic Press.

MacArthur, R. H., & E. J. Pianka. 1966. On optimal use of a patchy environment. *Am. Nat.* **100**:603-9.

MacArthur, R. M., & E. O. Wilson. 1967. *The theory of island biogeography.* Princeton: Princeton University Press.

Macevitz, S., & G. F. Oster. 1976. Modelling social insect populations. Part 2, Optimal reproductive strategies in annual eusocial insect colonies. *Behav. Ecol. Sociobiol.* **1**:265-82.

MacNair, M. R. 1978. An ESS for the sex ratio in animals, with particular reference to the social Hymenoptera. *J. Theor. Biol.* **70**:449–59.

Mangel, M., & C. W. Clark. 1988. *Dynamic modeling in behavioral ecology.* Princeton: Princeton University Press.

Margaria, R. 1976. *Biomechanics and energetics of muscular exercise.* Oxford: Clarendon Press.

Maynard Smith, J. 1974. The theory of games and the evolution of animal conflicts. *J. Theor. Biol.* **47**:209–21.

———. 1977. Parental investment: A prospective analysis. *Anim. Behav.* **25**:1–9.

———. 1979. Game theory and the evolution of behaviour. *Proc. Roy. Soc. B* **205**:475–88.

———. 1982. *Evolution and the theory of games.* Cambridge: Cambridge University Press.

Maynard Smith, J., & G. A. Parker. 1976. The logic of asymmetric contests. *Anim. Behav.* **24**:159–75.

Maynard Smith, J., & G. R. Price. 1973. The logic of animal conflict. *Nature* (London) **246**:15–18.

Minetti, A. E., & R. McN. Alexander. 1995. A theory of metabolic costs for bipedal gaits. *Physiol. Zool.* **68**(suppl.):45.

Mueller, U. G. 1991. Haplodiploidy and the evolution of facultative sex ratios in a primitively eusocial bee. *Science* **254**:442–44.

Parker, G. A. 1974a. Assessment strategy and the evolution of fighting behaviour. *J. Theor. Biol.* **47**:223–43.

———. 1974b. The reproductive behaviour and the nature of sexual selection in *Scatophaga sterocoraria* L. IX. Spatial distribution of fertilization rates and evolution of male search strategy within the reproductive area. *Evolution* **28**:93–108.

Parker, G. A., & J. Maynard Smith. 1990. Optimality theory in evolutionary biology. *Nature* **348**:27–33.

Parker, G. A., & R. A. Stuart. 1976. Animal behaviour as a strategy optimizer: Evolution of resource assessment strategies and optimal emigration thresholds. *Am. Nat.* **110**:1055–76.

Parker, G. A., & W. J. Sutherland. 1986. Ideal free distributions when individuals differ in competitive ability: Phenotype-limited ideal free models. *Anim. Behav.* **34**:1222–42.

Phillips, L. D. 1973. *Bayesian statistics for social scientists.* London: Nelson.

Pianka, E. R. 1976. Natural selection of optimal reproductive tactics. *Am. Zool.* **16**:775–84.

Rahn, H., C. V. Paganelli, & A. Ar. 1974. The avian egg: Air-cell gas tension, metabolism and incubation time. *Resp. Physiol.* **22**:297–309.

Rayner, J.M.V. 1985. Bounding and undulating flight in birds. *J. Theor. Biol.* **117**:47–77.

Reznick, D. A., H. Bryga, & J. A. Endler. 1990. Experimentally induced life-history evolution in a natural population. *Nature* **346**:357–59.

Saunders, P. T. 1980. *An introduction to catastrophe theory.* Cambridge: Cambridge University Press.

Schaffer, W. M. 1974. Selection for optimal life histories: The effects of age structure. *Ecology* **55**:291-303.

Sibly, R. 1975. How incentive and deficit determine feeding tendency. *Anim. Behav.* **23**:437-46.

Sibly, R. M., & P. Calow. 1986. *Physiological ecology of animals: An evolutionary approach.* Oxford: Blackwell.

Sibly, R., & D. McFarland. 1976. On the fitness of behaviour sequences. *Am. Nat.* **110**:601-17.

Snell, T. W., & C. E. King. 1977. Lifespan and fecundity patterns in rotifers: The cost of reproduction. *Evolution* **31**:882-90.

Snyder, A. W. 1977. Acuity of compound eyes: Physical limitations and design. *J. Comp. Physiol.* **116**:161-82.

Stearns, S. C. 1992. *The evolution of life histories.* Oxford: Oxford University Press.

Stephens, D. W., & J. R. Krebs. 1986. *Foraging theory.* Princeton: Princeton University Press.

Sterba, G. 1962. *Freshwater fishes of the world.* London: Vista Books.

Trivers, R. L., & H. Hare. 1976. Haplodiploidy and the evolution of the social insects. *Science* **191**:249-63.

Wainwright, S. A., W. D. Biggs, J. D. Currey, & J. M. Gosline. 1975. *Mechanical design in organisms.* London: Edward Arnold.

Webber, D. M., & R. K. O'Dor. 1986. Monitoring the metabolic rate and activity of free-swimming squid with telemetered jet pressure. *J. Exp. Biol.* **126**:205-24.

Weihs, D. 1974. Energetic advantages of burst swimming of fish. *J. Theor. Biol.* **48**:215-29.

Wersäll, J. 1956. Studies on the structure and innervation of the sensory epithelium of the cristae ampullares in the guinea pig. *Acta otolaryng* **126**(suppl.):1-85.

Woodcock, A., & M. Davis. 1980. *Catastrophe theory.* Harmondsworth: Penguin Books.

Zarrugh, M. Y., & C. W. Radcliffe. 1978. Predicting metabolic cost of level walking. *Europ. J. Appl. Physiol.* **38**:215-23.

Index